Understanding Pendulums

HISTORY OF MECHANISM AND MACHINE SCIENCE

Volume 12

Series Editor

MARCO GIOVANNI CECCARELLI

Aims and Scope of the Series

This book series aims to establish a well defined forum for Monographs and Proceedings on the History of Mechanism and Machine Science (MMS). The series publishes works that give an overviewof the historical developments, from the earliest times up to and including the recent past, of MMS in all its technical aspects.

This technical approach is an essential characteristic of the series. By discussing technical details and formulations and even reformulating those in terms of modern formalisms the possibility is created not only to track the historical technical developments but also to use past experiences in technical teaching and research today. In order to do so, the emphasis must be on technical aspects rather than a purely historical focus, although the latter has its place too.

Furthermore, the series will consider the republication of out-of-print older works with English translation and comments.

The book series is intended to collect technical views on historical developments of the broad field of MMS in a unique frame that can be seen in its totality as an Encyclopaedia of the History of MMS but with the additional purpose of archiving and teaching the History of MMS. Therefore the book series is intended not only for researchers of the History of Engineering but also for professionals and students who are interested in obtaining a clear perspective of the past for their future technical works. The books will be written in general by engineers but not only for engineers.

Prospective authors and editors can contact the series editor, Professor M. Ceccarelli, about future publications within the series at:

LARM: Laboratory of Robotics and Mechatronics
DiMSAT – University of Cassino
Via Di Biasio 43, 03043 Cassino (Fr)
Italy
E-mail: ceccarelli@unicas.it

For other titles published in this series, go to
http://www.springer.com/series/7481

L.P. Pook

Understanding Pendulums

A Brief Introduction

Springer

L.P. Pook
21 Woodside Road
Sevenoaks
TN13 3HF Kent
United Kingdom
les.pook@tesco.net

ISSN 1875-3442 e-ISSN 1875-3426
ISBN 978-94-007-3634-4 ISBN 978-94-007-1415-1 (eBook)
DOI 10.1007/978-94-007-1415-1
Springer Dordrecht Heidelberg London New York

Cover design: SPi Publisher Services

Printed on acid-free paper

Springer is part of Springer Science+Business Media (www.springer.com)

Preface

My interest in pendulums goes back to the late 1940s when I saw the Foucault pendulum at the Science Museum in London. Soon after we were married in 1960 my wife and I bought a longcase clock, which we still have. Keeping this, and other pendulum controlled clocks, running has taught me a lot about clock pendulums and the escapements used to control them, and aroused my interests in theoretical aspects of pendulums. For some years I have been collecting a broad range of information on pendulums, and this book is the result. Writing the book would not have been possible without consulting reference material held by the British Library in London, whose staff have been invariably helpful.

There are three related difficulties in writing about pendulums. The first is that it is an enormous subject so that a brief introductory book has to be selective in the material to be included. The second is that detailed explanations of pendulum behaviour often require advanced mathematics that can be difficult to follow, and may not be available for some situations. The third is that, although it is possible to define an ideal simple pendulum, physical pendulums inevitably deviate from an ideal, and it is difficult to assess the significance of these deviations. Despite their apparent simplicity, the behaviour of pendulums can be remarkably complicated. Historically, pendulums for specific purposes have been developed using a combination of simplified theory and trial and error.

Books on scientific subjects can be written at three different levels, firstly at a popular science level, with little or no significant mathematical content. Secondly, at an intermediate level with mathematical content at a recreation mathematics level, and used to illuminate the subject. Finally, written at an advanced level for specialists, with the mathematical content needed to understand the material at a serious mathematics level. There do not appear to be any introductory books on pendulums, written at an intermediate level, and covering a wide range of topics. This book aims to fill the gap. It is written for readers with some background in elementary geometry, algebra, trigonometry and calculus. Historical information, where available and useful for the understanding of various types of pendulum and their applications, is included. The wide range of topics covered means that the material is, to some extent, arranged in an arbitrary order. Extensive cross references are included so that individual chapters do not have to be read in order. Definitions are included in the index so that they can be easily located. Examples are given as case studies, sometimes with qualitative rather than quantitative explanations.

As described in Chap. 1, perhaps the best known use of pendulums is as the basis of clocks in which a pendulum controls the rate at which the clock runs. Interest in theoretical and practical aspects of pendulums, as applied to clocks, goes back more than four centuries. The concept of simple pendulums, which are idealised versions of real pendulums is introduced in Chap. 2, together with their analysis, and some variations on simple pendulums are described in Chap. 3. The application of pendulums to clocks is described in Chap. 4, with detailed discussion of the effect of inevitable differences between real pendulums and simple pendulums. In a clock, the objective is to ensure that the pendulum controls the timekeeping. However, pendulums are sometimes driven and how this affects their behaviour is described in Chap. 5. Pendulums are sometimes used for occult purposes. It is possible to explain some apparently occult results by using modern pendulum theory. For example, the chapter includes an explanation of why a ring suspended inside a wine glass, by a thread from a finger, eventually strikes the glass. Pendulums have a wide range of uses in scientific instruments, engineering, and entertainment. Some examples are given as case studies in Chaps. 6, 7 and 8.

January 2011 Les Pook

Contents

Notation

a	acceleration
a, b	Lissajous figure parameters, major and minor semi axes of ellipse
A	amplitude
A, B	arbitrary constants
A_1, A_2	arbitrary constants, amplitudes of successive peaks
B	force due to buoyancy
c	half diagonal of square
f	natural frequency, resonant frequency
f_p	frequency of periodic force
f_R	relative frequency
F	restoring force per unit distance
F_p	periodic force
$F(\beta)$	circular error at amplitude β
g	acceleration due to gravity, acceleration due to gravity at sea level
g'	acceleration due to gravity above sea level
g_e	effective value of g
$G(f)$	spectral density function
h	distance from frictionless pivot to centre of mass, a constant
H	height of point mass above its rest position
H_f	maximum value of H
I	moment of inertia, irregularity factor
j	distance from frictionless pivot to centre of oscillation
k	damping constant
K	kinetic energy
l	length of a simple pendulum, subscripts 1, 2 denote components of a double pendulum, effective length of a compound pendulum
l_V	vertical distance from pivot, virtual rod length
L	length, latitude
m	mass, subscripts 1, 2 denote components of a double pendulum
m_0, m_2, m_4	moments of spectral density function

m_V	velocity of point mass, m
M	mass
n	number of cycles, number of to and fro swings
N	number of balls in a Newton's cradle
p	momentum
$p(S)$	probability density
P	pendulum period $= T/2$, subscripts denote observed periods
$P(S)$	exceedance
Q	pendulum quality, quality factor
r	radius, radius of sphere, perpendicular distance from a pivot.
r, θ	polar coordinates
r, θ, ψ	spherical polar coordinates
r_s	radius of small circle
R	resisting force per unit velocity
RMS	root mean square
R_V	velocity of point R
$R(\tau)$	autocorrelation function
S	random process, positive peak, displacement from rest position, subscripts 1, 2 denote components of a double pendulum
SDF	spectral density function
S_m	Mean value of S
t	time
T	time of swing of a pendulum, period of simple harmonic motion or damped harmonic motion, time of one rotation about an axis, wave passing period, total time
T_E	effective time of rotation
V	velocity, strike velocity
W	potential energy
W_a	absorbed energy
x	distance from a fixed point, displacement from rest position
x, y	Cartesian coordinates, subscripts 1, 2 denote displacements of a double pendulum
X	pivot displacement
α	ratio between periods
$1/a$	relative frequency
α, ε	arbitrary constants
β	angle of swing of pendulum from rest position, maximum value of θ, angle of fall
γ	angular interval, angle of rise
δ	logarithmic decrement
ε	phase, spectral bandwidth
$\varepsilon_1, \varepsilon_2$	arbitrary constants
κ	radius of gyration
κ_m	radius of gyration about centre of mass
ϕ	phase angle, root mean square value of S

ϕ^2	mean square value of S
σ	standard deviation
σ^2	variance
τ	time constant, increment of time
θ	pendulum angle, subscripts 1, 2 denote components of a double pendulum, angle of rotation
ω	angular velocity (angles in radians)
ω_d	damped angular velocity
ω_p	angular velocity of periodic force

About the Author

Leslie Philip (Les) Pook was born in Middlesex, England in 1935. He obtained a BSc in metallurgy from the University of London in 1956. He started his career at Hawker Siddeley Aviation Ltd, Coventry in 1956. In 1963 he moved to the National Engineering Laboratory, East Kilbride, Glasgow. In 1969, while at the National Engineering Laboratory, he obtained a Ph.D. in mechanical engineering from the University of Strathclyde. Dr. Pook moved to University College London in 1990. He retired formally in 1998 but remained professionally active in the fields of metal fatigue and fracture mechanics, and was a visiting professor at University College London until 2009. He now has more tine to pursue long standing interests in recreational mathematics, including flexagons, and in horology, especially synchronous electric clocks. He is a Fellow of the Institution of Mechanical Engineers, a Fellow of the Institute of Materials, Minerals and Mining, and a Fellow of the European Structural Integrity Society. Les married his wife Ann in 1960. They have a daughter, Stephanie, and a son, Adrian.

Chapter 1
Introduction

Perhaps the best known use of pendulums is as the basis of clocks in which a pendulum controls the rate at which the clock runs. Many clocks have visible pendulums; a wall clock with a glazed door over the dial and pendulum is shown in Fig. 1.1. Indeed, the association of pendulums with clocks is so pervasive that, since the advent of quartz clocks, it is possible to buy clocks with visible swinging pendulums that are independent of the timekeeping of the clock.

A clock pendulum usually consists of a relatively light *pendulum rod* with a heavy *bob* at the lower end (Fig. 1.1). The upper end is attached to a suitable mounting by a *suspension* which allows the pendulum to swing freely. Traditionally, the suspension of the pendulum in a clock is a flat spring. For example, Fig. 1.2a shows the suspension spring and the upper end of the pendulum rod in the movement of an eighteenth century longcase clock. A flat *spring* is still used to suspend the pendulum in most pendulum clocks. The bob and the lower part of the pendulum are shown in Fig. 1.2b. The *rating nut* below the bob is used to adjust the length of the pendulum. The pendulum is connected to an *escapement* which controls the clock by allowing one tooth of the '*scape wheel* to escape for each swing of the pendulum. The escapement of the eighteenth century clock is an *anchor escapement* (Fig. 1.2c), a reference to its shape, A sketch of an anchor escapement is shown in Fig. 1.3. The specially shaped *pallets* at the ends of the anchor engage with specially shaped teeth on the 'scape wheel. The anchor is mounted on an *arbor*, which is the clockmaker's name for an axle, and is connected to the pendulum by a *crutch* (Fig. 1.2a). The 'scape wheel has 30 teeth which means that, for a seconds pendulum it turns once per minute so its arbor can be used to drive a seconds hand.

Interest in the behaviour of pendulums, in particular calculation of times of swing, dates back to the observation, usually attributed to Galileo, that the time of swing of a pendulum is independent of the amplitude of its oscillations (Edwardes 1977). In other words the pendulum is *isochronous*. The story is that in 1581 Galileo was sitting in Pisa Cathedral, and compared the times of oscillations of suspended lamps with the pulse in his wrist.

L.P. Pook, *Understanding Pendulums: A Brief Introduction*, History of Mechanism and Machine Science 12, DOI 10.1007/978-94-007-1415-1_1,

Fig. 1.1 A pendulum controlled wall clock with glazed door over the dial and pendulum

The seminal book on theoretical and practical aspects of pendulums, as applied to clocks, is 'Horologium' by Christiaan Huygens (Huygens 1658/1977). 'Horologium oscillatorium. The pendulum clock or geometrical demonstrations concerning the motion of pendula as applied to clocks'. (Huygens 1673/1986) is an updated and extended version. Both books were published in Latin. The references cited are modern translations. Historical surveys from a horological perspective include those by Edwardes (1977) and Roberts (2003, 2004). The book by Matthys (2004) is a modern account of accurate clock pendulums, and Britten (1978) includes a short general survey of clock pendulums. Some other applications of pendulums are included in Dunkerley (1910), Lamb (1923) and Tobin (2003). Baker and Blackburn (2005) is an advanced physics textbook on a wide range of pendulums. Matthews (2000) is an historical survey from a pedagogical perspective.

Scientific applications of pendulums include the *Foucault pendulum* used to demonstrate rotation of the Earth (Tobin 2003), and the *Charpy impact test*, used to measure the fracture resistance of metals (Siewert et al. 2000). Industrial applications include the *Watt steam governor* used to regulate steam engines (Lineham 1914) and the *tension leg platforms* used for the offshore recovery of

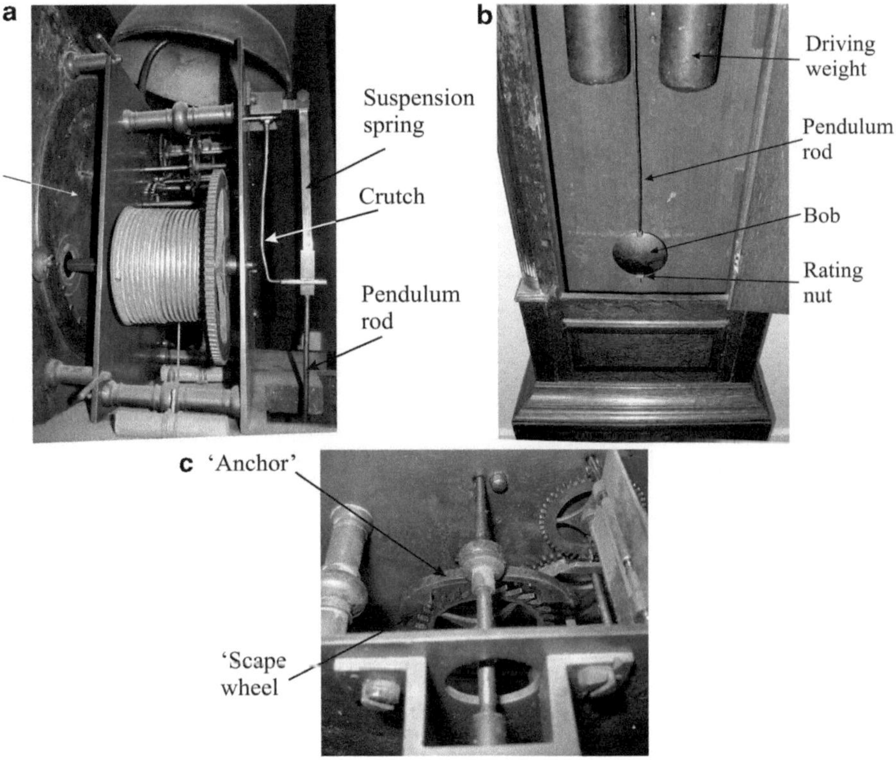

Fig. 1.2 Eighteenth century longcase clock. (**a**) Movement. (**b**) Pendulum rod and bob. (**c**) Anchor escapement

Fig. 1.3 Sketch of anchor escapement

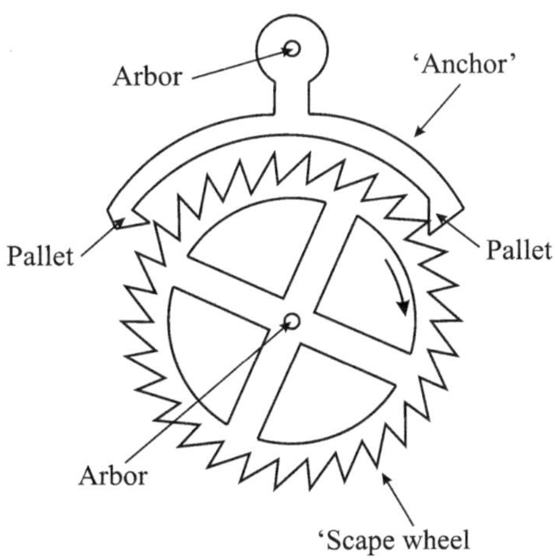

oil and gas (Patel and Witz 1991). Recreational uses include *children's swings* (Case and Swanson 1990), and *pendulum harmonographs* used to draw figures known as *harmonograms* (Whitaker 2001). Pendulums are sometimes used for occult purposes (Jurriaanse 1987).

A pendulum is a simple system, but its behaviour can be remarkably complex (Nelson and Olsson 1986). Detailed understanding requires advanced mathematics that can be difficult to follow. Fortunately, the results of analyses are usually easy to understand, and the derivations given in this book can usually be omitted on a first reading. Simplifications are often used in order to make mathematical analysis more tractable. Whether particular simplifications are acceptable depends largely on the required accuracy of numerical results.

For the four centuries following Huygens' seminal work pendulum behaviour was regarded as essentially regular. The important properties of a pendulum were assumed to be its *time of swing*, and the variability of the time of swing. Much effort was devoted to theoretical and practical aspects of improving the accuracy of pendulums used to control clocks. Pendulum controlled clocks can be remarkably accurate. Thus, a physicist might be satisfied with an accuracy of 1% (Parwani 2004), whereas a horologist might require a clock to have an accuracy of 1 s per day, that is about 0.001%, or even better (Matthys 2004). An accuracy of 1 ms per day (about 0.000001%) is achievable (Feinstein 1955; Roberts 2003). For domestic purposes an accuracy of 2 min per week (0.02%) is usually acceptable.

More recently it has been appreciated that pendulum behaviour can be irregular. This means that *chaos* theory is needed for understanding of pendulum behaviour. In other words a pendulum can be a *chaotic system*. In scientific terminology chaos is shorthand for *chaotic dynamics* and therefore has a different meaning from its everyday meaning of utter confusion. Chaotic dynamics was originated in 1913 by Henri Poincaré (Baker and Blackburn 2005). It now has an extensive literature, with several textbooks devoted to the topic, for example Baker and Gollub (1996). Burger and Starbird (2005) include an elementary introduction to chaos. The introduction by Hall (1992) includes some of the history of chaos.

Chaotic dynamics is used to describe the *chaotic behaviour* of a chaotic system. Chaotic behaviour is not random. If the initial conditions are known with sufficient accuracy then the subsequent behaviour can, in principle, be predicted precisely. In practice, in a chaotic system the subsequent behaviour is so sensitive to the initial conditions that an attempt to reproduce the initial conditions is never sufficiently precise, and the outcome on successive attempts can vary widely.

Chaotic behaviour can only appear in *nonlinear systems* (Baker and Gollub 1996). There is no generally accepted precise definition of a nonlinear system. Broadly speaking it is a system whose behaviour cannot be described by relatively simple equations. Conversely, a *linear system* is one whose behaviour can be described by relatively simple equations. However, the definition could equally well be reversed by stating that a nonlinear system is one in which chaotic behaviour can occur. The complicated equations needed to describe its behaviour means that a pendulum is a nonlinear system.

As an everyday example of chaotic behaviour, once a die is thrown which of the 6 numbers appears uppermost when it comes to rest is precisely determined, but it cannot be thrown with sufficient precision to ensure that a particular number appears uppermost. Chaos must be distinguished from *probability*. For example, the probability of a particular number appearing when a die is thrown is 1/6.

References

Baker GL, Blackburn JA (2005) The pendulum. A case study in physics. Oxford University Press, Oxford

Baker GL, Gollub JP (1996) Chaotic dynamics. An introduction, 2nd edn. Cambridge University Press, Cambridge

Britten FJ (1978) The watch & clock makers' handbook, dictionary and guide, 16th edn. Arco Publishing Company, New York (Revised by Good R)

Burger EB, Starbird M (2005) Coincidences, chaos and all that math jazz: making light of weighty ideas. W W Norton & Company, New York

Case WB, Swanson MA (1990) The pumping of a swing from the seated position. Am J Phys 58(5):463–467

Dunkerley S (1910) Mechanisms, 3rd edn. Longmans, Green and Co., London

Edwardes EL (1977) The story of the pendulum clock. John Sherratt & Son, Altringham

Feinstein GF (1995) M Fedchenko and his pendulum astronomical clocks. NAWCC Bull 1995:169–184

Hall N (ed) (1992) The New Scientist guide to chaos. Penguin, London

Huygens C (1658/1977) The horologium. Latin facsimile and English translation. In: Edwardes EL. The story of the pendulum clock. John Sherratt & Son, Altringham, pp 60–97

Huygens C (1673/1986) Horologium oscillatorium. The pendulum clock or geometrical demonstrations concerning the motion of pendula as applied to clocks. The Iowa State University Press, Iowa (Trans by Blackwell RJ)

Jurriaanse D (1987) The practical pendulum book, with instructions for use and thirty-eight pendulum charts. The Aquarium Press, Wellingborough

Lamb H (1923) Dynamics, 2nd edn. Cambridge University Press, Cambridge

Lineham WJ (1914) A textbook of mechanical engineering. 11th edn. Chapman and Hall, London

Matthews RJ (2000) Time for science education. How teaching the history and philosophy of pendulum motion can contribute to science literacy. Kluwer Academic/Plenum Publishers, New York

Matthys R (2004) Accurate clock pendulums. Oxford University Press, Oxford

Nelson RA, Olsson MG (1986) The pendulum – rich physics from a simple system. Am J Phys 54(2):112–121

Parwani RR (2004) An approximate expression for the large angle period of a simple pendulum. Eur J Phys 25(1):37–39

Patel MH and Witz JA (1991) Compliant offshore structures. Butterworth-Heinemann, Oxford

Roberts D (2003) Precision pendulum clocks: 300 year quest for accurate timekeeping in England. Schiffer Publishing, Atglen

Roberts D (2004) Precision pendulum clocks: France, Germany, America, and recent advancements. Schiffer Publishing, Atglen

Siewert TA, Manahan MP, McCowan CN, Holt JH, Marsh FJ, and Ruth EA (2000) The history and importance of impact testing. In: Siewert TA, Manahan MP (eds) Pendulum impact testing: a century of progress. American Society for Testing and Materials, West Conshohocken, pp 3–16 (STP 1380)

Tobin W (2003) The life and science of Léon Foucault. The man who proved that the earth rotates. Cambridge University Press, Cambridge

Whitaker RJ (2001) Harmonographs I. Pendulum design. Am J Phys 69(2):162–173

Chapter 2
Simple Pendulums

2.1 Introduction

Ideal pendulums are idealisations of *real pendulums* that are often used in order to make mathematical analysis more tractable. Ideal pendulums are not physically realistic. In scientific terminology, an ideal pendulum is a *model* of a real pendulum. Results of analyses of ideal pendulums often provide useful approximations for real pendulums, such as the clock pendulums shown in Figs. 1.1 and 1.2b. A *simple pendulum* is a particular type of ideal pendulum, and the term is often used without explanation. In a simple pendulum, the bob of a real clock pendulum is replaced by a *particle* whose dimensions are small enough for it to be adequately represented by a mathematical point (Lamb 1923). When a specified mass, m, is ascribed to a particle it is called a *point mass*. The only force on a simple pendulum is that due to *gravity* on the point mass.

There are actually two types of simple pendulum (Lamb 1923). In a *simple rod pendulum* (Fig. 2.1a) the rod of a real clock pendulum (Figs. 1.1 and 1.2b) is replaced by a rigid, massless rod, length l, with a point mass m, replacing the bob, attached to the lower end. *Rigid* means that the rod does not deform under an applied load, in particular the length always stays the same. In other words the rod is *inextensible*. The suspension of a real pendulum, such as the suspension spring in Fig. 1.2a, is replaced by a horizontal *frictionless pivot*, where *frictionless* means that the pivot does not exert any force as the rod rotates about the pivot. In the figure the pivot is perpendicular to the paper so that the rod, with the point mass attached, swings in a vertical plane. The frictionless pivot is the *suspension point*. This is fixed in space so the path of the point mass is a circular arc, radius l. It is usually sufficient to regard a point on the Earth's surface as being fixed in space. The simple rod pendulum has one *degree of freedom*. In other words, one parameter is needed to describe the position of the point mass, m. An example of an appropriate parameter is the *pendulum angle*, θ, in Fig. 2.2. A simple rod pendulum is not a chaotic system, so it does not display chaotic behaviour.

L.P. Pook, *Understanding Pendulums: A Brief Introduction*, History of Mechanism and Machine Science 12, DOI 10.1007/978-94-007-1415-1_2,

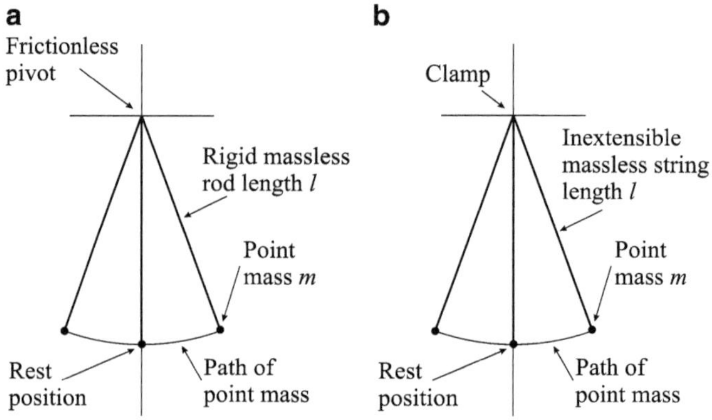

Fig. 2.1 Simple pendulums. (**a**) Simple rod pendulum. (**b**) Simple string pendulum

Fig. 2.2 Resolved forces on the point mass of a simple rod pendulum

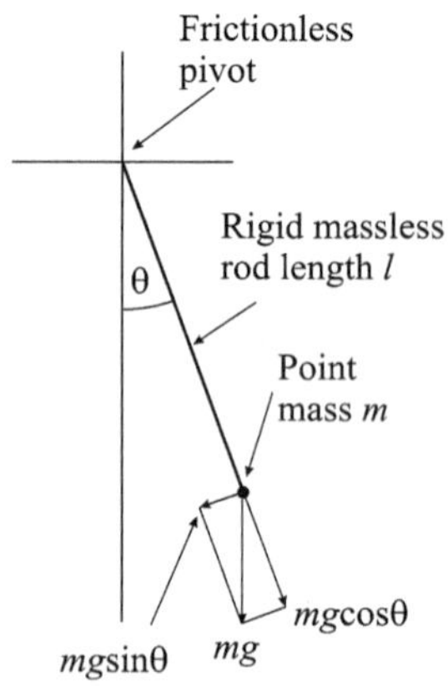

In a *simple string pendulum* (Fig. 2.1b) the string of a real pendulum, such as the pendulum for occult uses shown in Fig. 2.3, is replaced by a massless string, length l, with a point mass m, replacing the bob, attached to the lower end. The string is inextensible in the sense that, when pulled taut, it is always the same length. The string has no resistance to bending and the suspension point is a clamp at the upper end. The clamp is fixed in space so, provided that the string remains taut, the path of the point mass is on a sphere, radius l. The simple string pendulum with a taut string has two degrees of freedom so two parameters are needed to describe the position of the point mass. The path taken by the point mass depends on how the pendulum is launched. A simple string pendulum is a chaotic system that can display chaotic behaviour.

Gravity acting on the point mass of a simple pendulum is assumed to be uniform and to act vertically downwards. This is the *flat Earth assumption*, and is not quite correct for a real pendulum because the direction of gravity is towards the centre of the Earth. Hence its direction changes very slightly for different positions of the

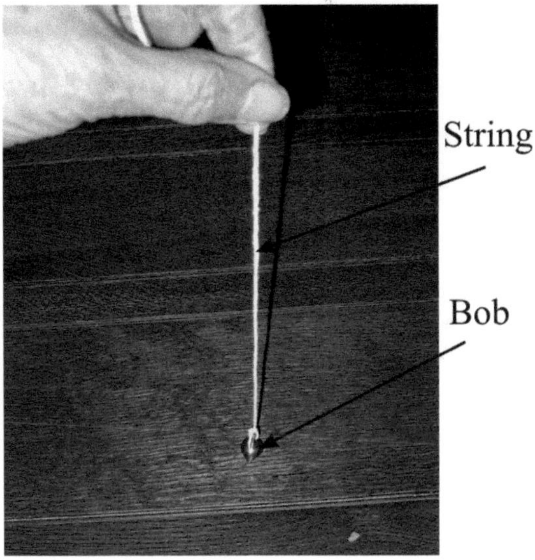

Fig. 2.3 A pendulum for occult uses

pendulum bob (Rawlings 1993; Burko 2003; Matthys 2003). Simple pendulums are *ideal pendulums* and, once launched, they will continue to swing indefinitely with the same amplitude.

Simple pendulums are approximately isochronous for small amplitudes, and the motion of the point mass approximates to *simple harmonic motion*, which is described below. Both types of simple pendulum are in *stable equilibrium* when in the *rest position*, which is vertically downwards. Stable equilibrium means that, following a small displacement, a simple pendulum tends to return to the rest position. If the simple rod pendulum is vertically upwards then it is in *unstable equilibrium* and, following a small displacement, the point mass does not return to its initial position. A *simple string pendulum* does not have an unstable equilibrium position.

For some purposes it is not necessary to distinguish between the two types of simple pendulum, and authors often refer to 'the *simple pendulum*' without explanation. Textbook definitions are usually of only one type. For example, Whelan and Hodgson (1977) include a sketch of what they call a simple pendulum with a legend 'light string' so it is a simple string pendulum, and Baker and Blackburn (2005) define a simple pendulum as 'an infinitely light rod of length, l, that is itself attached to a frictionless pivot point', so it is a simple rod pendulum.

Mathematically, a point mass which is a *bead* sliding on a frictionless circular wire lying in a vertical plane is equivalent to a *simple rod pendulum* (Lamb 1923). This is the *bead on wire analogy*. Similarly, a point mass sliding in a frictionless spherical bowl is equivalent to a *simple string pendulum*. This is the *point mass in sphere analogy*. It can be approximated by a small ball rolling in a bowl (Fig. 2.4).

Fig. 2.4 Ball bearing in a turned wooden bowl

2.2 Simple Harmonic Motion

The motion of a *simple rod pendulum* is an oscillation that, for small amplitudes, approximates to *simple harmonic motion*. This arises in systems with one degree of freedom where the *restoring force*, F, towards a *rest position* is proportional to the distance from the rest position. A simple example is a point mass, m, which can move along a straight line with a restoring force towards a fixed point, O, that is proportional to the distance from the fixed point, x. Simple harmonic motion is usually regarded as a linear system.

If the force at unit distance, taken as positive, is F, then the force at distance x is $-Fx$, the sign is always the opposite of that of x. The corresponding differential equation is

$$m\frac{\mathrm{d}^2x}{\mathrm{d}y^2} = -Fx \tag{2.1}$$

where t is time. The solution is (Lamb 1923)

$$X = A\cos(nt) + B\sin(nt) \tag{2.2}$$

or

$$x = a\cos(\omega t + \varepsilon) \tag{2.3}$$

where

$$n^2 = F/m \tag{2.4}$$

and the *angular velocity*, ω, is given by

$$\omega = \sqrt{\frac{F}{m}} \tag{2.5}$$

A, B, or α, ε are arbitrary constants.

The motion is an oscillation in which particular values of x (amplitude) and $\mathrm{d}x/\mathrm{d}t$ (velocity) recur at intervals of nt of 2π. The *period*, T, of these intervals is given by

$$T = \frac{2\pi}{\omega} = 2\pi\sqrt{\frac{m}{F}} \tag{2.6}$$

The period is independent of the initial conditions, so the oscillation is isochronous. In particular, the period is independent of the amplitude of the oscillations. The reciprocal of T is the *natural frequency*, f, and is given by

$$f = \frac{1}{2\pi}\sqrt{\frac{F}{m}} \tag{2.7}$$

Because it oscillates at its natural frequency simple harmonic motion is, in physics terminology, a *resonant system*. Differentiating Eq. 2.6, the velocity of the point mass, $\mathrm{d}x/\mathrm{d}t$, is given by

$$\frac{\mathrm{d}x}{\mathrm{d}t} = \frac{-Fx}{m} \tag{2.8}$$

Re-arranging Eq. 2.1 the acceleration is given by

$$\frac{\mathrm{d}^2x}{\mathrm{d}t^2} = \frac{-Fx}{m} \tag{2.9}$$

A feature of simple harmonic motion is that a plot of the displacement of the point mass, m, against time is a sine wave which repeats indefinitely (Fig. 2.5). Another graphical representation of simple harmonic motion is shown in Fig. 2.6. In the figure a point R moves around a circle with constant angular velocity, ω, as defined by Eq. 2.5. The vector OR has the same angular velocity. When it is at point R the projection, point S, onto the diameter of the circle, AB, gives the position of the point mass moving with simple harmonic motion along the diameter. The radius of the circle, OA, is the amplitude of the oscillation, A, and ωRS is the velocity of the point mass at point S. The angle AOR is the *phase* of the point mass at S.

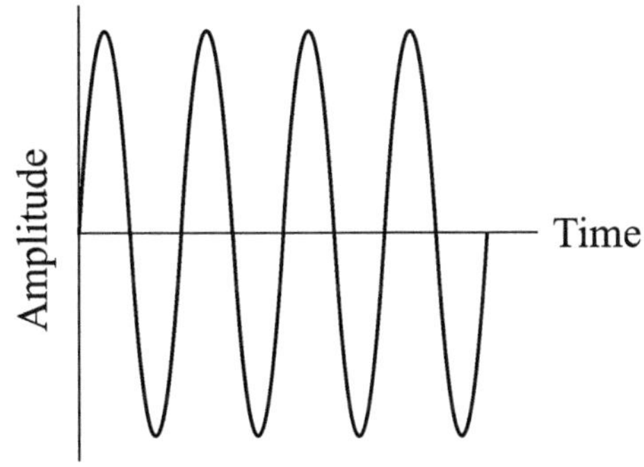

Fig. 2.5 Simple harmonic motion. The sine wave repeats indefinitely

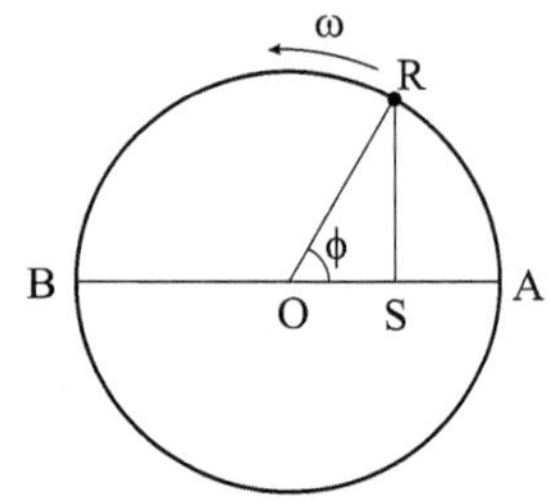

Fig. 2.6 Graphical representation of simple harmonic motion and Lissajous figures

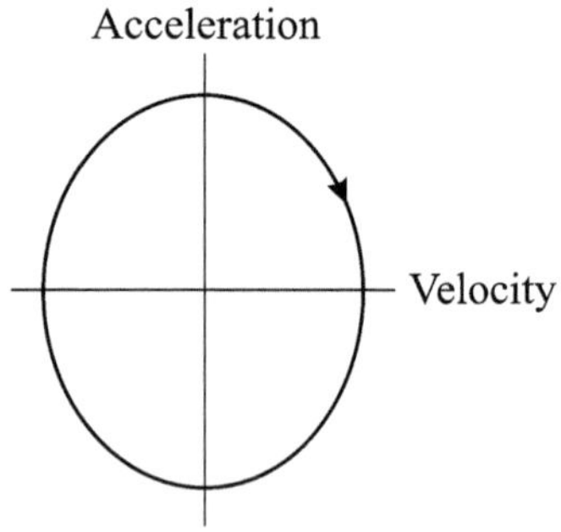

Fig. 2.7 Phase diagram for simple harmonic motion. The *arrow* shows the sense of increasing time

In terms of the amplitude, A, angular velocity, ω, and time, t, the velocity, V, of the point mass is given by (Bateman 1977)

$$V = \frac{dx}{dt} = -A\omega \sin(\omega t) \tag{2.10}$$

where x is the displacement from the rest position, and the negative sign indicates that the point mass is moving towards the rest position. As is intuitively obvious, the equation shows that velocity has arithmetic maxima as the point mass passes through the rest position, and is zero when the point mass is furthest from the rest position (points A and B on Fig. 2.6).

The acceleration, α, is given by

$$\alpha = \frac{d^2x}{dt^2} = -\omega^2 x \tag{2.11}$$

where the negative sign indicates that acceleration of the point mass is towards the rest position. The acceleration is zero as the point mass passes through the rest position, and has arithmetic maxima when the point mass is furthest from the rest position. In a *phase diagram* acceleration is plotted against velocity. The phase diagram for simple harmonic motion is an ellipse (Fig. 2.7). One complete circuit of the phase diagram, in the direction shown by the arrow, represents one complete oscillation of the point mass.

2.3 Analysis of a Simple Rod Pendulum

In a *simple rod pendulum*, as defined in Sect. 2.1, the path of the point mass, m, is a circular arc, with centre at the frictionless pivot (Fig. 2.1a). The motion of the pendulum is within a plane so it is a *planar mode of oscillation*. The behaviour of a simple rod pendulum can be described precisely by relatively simple equations, and it does not display chaotic behaviour (Sect. 2.1). For small amplitudes the motion of the point mass approximates to simple harmonic motion (previous section), and physics textbooks usually include an analysis, for example Whelan and Hodgson (1977), and Baker and Blackburn (2005). These analyses relay on the approximation that, for small values of the pendulum angle (Fig. 2.2), $\sin\theta \approx \theta$ where θ is in radians. A typical analysis is given below. This is similar to the analysis of simple harmonic motion.

A force mg acts vertically downwards on the point mass of a simple rod pendulum, where m is its mass, and g is the acceleration due to gravity. Because it is *vector* with both a magnitude and a direction the force mg can be resolved into a component parallel to the instantaneous direction of travel (*tangential force*) and perpendicular to the instantaneous direction of travel (*normal force*). For a pendulum deflected through an angle θ, in radians, from the rest position these are $mg\sin\theta$ and $mg\cos\theta$ respectively (Fig. 2.2). The normal force $mg\cos\theta$ is reacted by a force of opposite sign in the rod and has no effect on the behaviour of the pendulum. It can therefore be neglected in the calculation of the *time of swing*, T. The tangential force $mg\sin\theta$ acts to restore the point mass towards its rest position, so is the restoring force. For small θ, $\sin\theta \approx \theta$, where θ is in radians, so the restoring force is approximately $mg\theta$.

The displacement, S, of the point mass along the arc during a deflection of the pendulum through an angle θ from the rest position is $l\theta$, where l is the length of the rod. Hence, F is approximately proportional to S, which implies that the motion of the point mass is an oscillation approximating to simple harmonic motion (previous section). The position and velocity of the point mass repeats exactly at the time interval, T, which is the time of swing. Simple harmonic motion is a linear system (previous section). Making the approximation $\sin\theta \approx \theta$ means that the equations of motion of a pendulum have been *linearised* in order to make the mathematics more tractable.

The force per unit distance is given approximately by

$$F = \frac{mg\theta}{l\theta} = \frac{mg}{l} \tag{2.12}$$

Substituting the value of F given by Eq. 2.12 into Eq. 2.6 leads to

$$T = 2\pi\sqrt{\frac{l}{g}} \tag{2.13}$$

This well known equation for the approximate time of swing of small amplitude oscillations of a *simple rod pendulum* was first formulated by Huygens (1673/1986). The time of swing, T, given by the equation is sometimes called the *natural period* of the pendulum. The reciprocal of T is the *natural frequency*, f, of the pendulum and is given by

$$f = \frac{1}{2\pi}\sqrt{\frac{g}{l}} \tag{2.14}$$

Because it oscillates at its natural frequency a pendulum is a resonant system. For clocks the *pendulum period*, P, is usually taken as the time between successive ticks and $P = T/2$. Hence Eq. 2.13 becomes

$$P = \pi\sqrt{\frac{l}{g}} \tag{2.15}$$

2.3.1 Acceleration Due to Gravity

The value of the acceleration due to *gravity*, g, varies with location on the Earth. It decreases from the Equator to the Poles, and decreases slightly with height above sea level. There are also small local variations. From experimental data the value of g at sea level is given by (Lamb 1923)

$$g = 9.7803(1 + 0.0053\sin^2\phi) \tag{2.16}$$

where ϕ is the latitude and g is in m/s^2. From the Equator to a pole the value of g increases by just over 1/2 per cent. At latitude 45° $g = 9.8062\,\text{m/s}^2$. The value of g at above sea level, g', is given by

$$g' = g\,(1 - 0.000\,0003h) \tag{2.17}$$

where h is the height above sea level in metres. For most purposes the difference between g' and g is negligible.

Inserting $g = 9.81\,\text{m/s}^2$ into Eq. 2.13 gives the lengths of *simple rod pendulums* for various *pendulum periods* shown in Table 2.1.

The length of a simple rod pendulum with a period of 1 s is approximately 1 m. An early proposed definition of the metre was that it should be the length of a seconds pendulum (Matthews 2000). The difficulty was that g is not constant over the Earth's surface so it would be necessary to specify a location, with the

Table 2.1 Periods and lengths for a simple rod pendulum

Period, P (s)	0.5	1	1.25	1.5	2
Length, l (m)	0.2485	0.994	1.553	2.236	3.975

concomitant political problems. A standard brass metre was adopted in 1799. The current definition of the metre, adopted in 1983, is 'the length of path travelled by light in vacuum during a time interval of 1/299 792 458 of a second'. In turn the second is based on a natural frequency of a Caesium atom, which is 9 192 631 770 Hz (Baker and Blackburn 2005).

2.3.2 Accelerations of a Simple Rod Pendulum

Acceleration is a vector and when a particle is moving along a prescribed curve it is often convenient to resolve its acceleration into tangential and normal components. Equation (2.11) shows that, for simple harmonic motion of a point mass along a straight line, accelerations are always along this line towards the rest position of the point mass. As the point mass passes the rest position the velocity is an arithmetic maximum (Eq. 2.10) and the acceleration is zero. At the ends of the line along which the point mass is moving the velocity is zero and the accelerations are arithmetic maxima. The point mass of a *simple rod pendulum* (Fig. 2.1a) moves along a prescribed curve, which is a circular arc. The tangential components of accelerations are along the circular arc, and are similar to the accelerations for simple harmonic motion along a straight line. In particular, as the point mass passes the rest position the velocity is an arithmetic maximum, and the tangential component of acceleration is zero. At the ends of the arc along which the point mass is moving the velocity is zero, and the tangential components of accelerations are arithmetic maxima.

In a simple rod pendulum, except at the ends of the arc, there is a normal component of acceleration due to *centripetal force* towards the frictionless pivot. This normal acceleration, α, is given by (Kreyszig 1983)

$$|a| = \omega^2 r \tag{2.18}$$

where r is the radius of the arc, that is the pendulum length l. This acceleration has maxima when the point mass passes through the rest position, and is zero at the ends of the arc. The resulting centripetal force is ma where m is the mass of the point mass. It is sometimes stated, erroneously, that the force $mg\cos\theta$ (previous section) is cancelled out by the force in the rod (Camponario 2006). This neglects centripetal forces.

Combining the components of accelerations for the point mass of a simple rod pendulum leads to the resultant *acceleration directions* shown schematically in Fig. 2.8a. As the point mass passes the rest position acceleration is vertically upwards. At the ends of the arc accelerations of the point mass are tangential towards the rest position. At intermediate positions resultant acceleration directions are at intermediate angles. These acceleration directions are notoriously difficult to visualise (Matthews 2000). Directions for tangential acceleration components are shown in Fig. 2.8b.

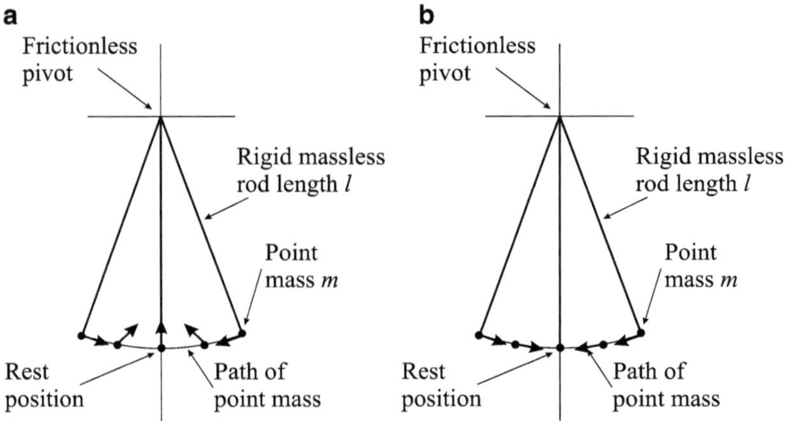

Fig. 2.8 (**a**) Acceleration directions (*arrows*) for the point mass of a simple rod pendulum. (**b**) Tangential component

Fig. 2.9 Phase diagram for a simple rod pendulum. The *arrow* shows the sense of time

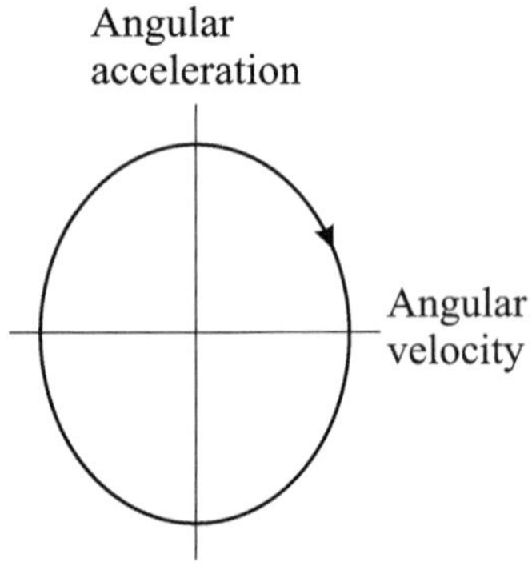

Sketching the acceleration directions for a simple rod pendulum is sometimes asked as a trick question in which it is not made clear whether or not the normal (centripetal) component should be included. If normal components are included then Fig. 2.8a is correct. If only accelerations that affect motion of the point mass are required then Fig. 2.8b is correct.

The position of a simple rod pendulum is completely defined by the angle θ (Fig. 2.2) and it is sometimes convenient to refer to its *angular velocity* and *angular acceleration*. The relationship between angular velocity and angular acceleration can be represented by a *phase diagram* (Fig. 2.9, cf. Fig. 2.7) in which angular acceleration is plotted against angular velocity. One complete circuit of the phase diagram, in the direction shown by the arrow, represents one complete oscillation of the simple rod pendulum. Angular velocity is in radians so the frequency is the angular velocity divided by 2π. For small oscillations of a simple rod pendulum the phase diagram is approximately an ellipse.

2.3.3 *Potential and Kinetic Energy of a Simple Rod Pendulum*

A *simple rod pendulum* is an ideal pendulum that, once launched, will continue to swing indefinitely. In physics terminology it is a *non dissipative system* in which the total energy is constant. The total energy can be resolved into *potential energy* and *kinetic energy*. The potential energy, W, of a simple rod pendulum with a point mass, m, (Fig. 2.1a) is

$$W = mgH = mgl\,(1 - \cos\theta) \tag{2.19}$$

where g is the acceleration due to gravity, H is the height of the point mass, m, above the rest position, l is the length of the pendulum, and θ is the pendulum angle (Fig. 2.2). The potential energy has maxima at the ends of the arc where H has its maximum value, H_{f}, and is zero when the point mass, m, passes through the rest position. The kinetic energy, K, is

$$K = \frac{1}{2}mV^2 \tag{2.20}$$

where V is the velocity of the point mass, m. The kinetic energy is zero at the ends of the arc, and has maxima when the point mass, m, passes through the rest position. Combining Eqs. 2.19 and 2.20, the velocity of the point mass as it passes through the rest position is

$$V = \sqrt{2gl\,(1 - \cos\beta)} \tag{2.21}$$

where β is the *angle of swing*, that is value of θ at the end of the arc.

2.3.4 *Circular Error of a Simple Rod Pendulum*

For large amplitudes a *simple rod pendulum* is not isochronous. The difference between the period of a simple rod pendulum given by Eq. 2.15 and the actual period is known is the *circular error* (Britten 1978). This is a misnomer, but it is the accepted horological term. Circular error was first derived theoretically by Huygens (1673/1986), using geometric methods. Modern derivations make use of an elliptic integral (Loney 1913; Lamb 1923; Feinstein 1995; Baker and Blackburn 2005). Expanding the elliptic integral as an infinite series the pendulum period, P, is given by

$$P = \pi\sqrt{\frac{l}{g}}\left[1 + \frac{\beta^2}{16} + 11\frac{\beta^4}{3072} + 519\,\frac{\beta^6}{2211840}\cdots\right] \tag{2.22}$$

where β is the angle of swing in radians, measured from the rest position. As $\beta \rightarrow \pi$ $P \rightarrow \infty$. If β is small the first term is sufficient and

$$P = \pi \sqrt{\frac{l}{g}} \left[1 + \frac{\beta^2}{16} \right] \tag{2.23}$$

Various approximations for large values of β have been suggested. For example, the expression (Parwani 2004)

$$P = \pi \sqrt{\frac{l}{g}} \left[\frac{\sin\left(\sqrt{3}\,\beta/2\right)}{\left(\sqrt{3}\,\beta/2\right)} \right]^{-1/2} \tag{2.24}$$

is within 1% for β up to 2.25 rad (129°). Other expressions are given by Johannessen (2010) and Berléndaz et al. (2010).

Equation (2.24) can be written as

$$P = \pi \sqrt{\frac{l}{g}} F(\beta) \tag{2.25}$$

where $F(\beta)$ is the circular error at amplitude β. Some values of $F(\beta)$ are shown in Table 2.2 (Lamb 1923). The circular error for small amplitudes is shown in Fig. 2.10 in terms of seconds per day.

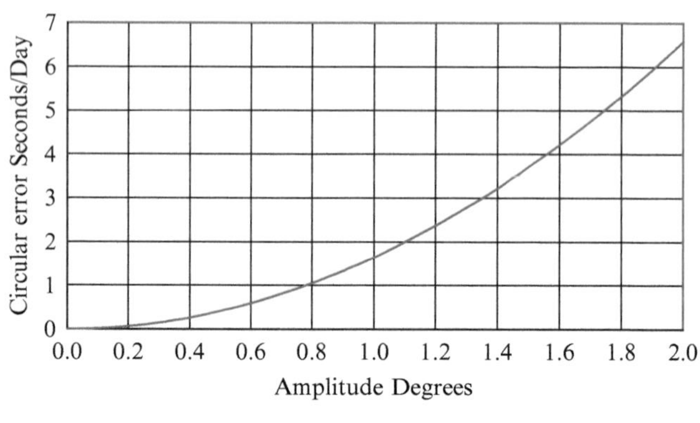

Fig. 2.10 Circular error of a simple rod pendulum

Table 2.2 Circular error of a simple rod pendulum

Amplitude degrees	$F(\beta)$	Amplitude degrees	$F(\beta)$	Amplitude degrees	$F(\beta)$
0	1.000000	5	1.000476	60	1.073182
1	1.000019	10	1.001907	90	1.180340
2	1.000076	20	1.007669	120	1.372880
3	1.000170	30	1.017409	150	1.762204
4	1.000305	45	1.039973	180	∞

If a *simple rod pendulum* is launched with a sufficient initial angular velocity then it can make *continuous rotations* about its frictionless pivot. The path of the point mass is a vertical circle, and can be in either direction around the circle. By analogy with planetary motions the path travelled by the point mass is an *orbit.* The angular velocity, ω, has a maximum at the lowest point of the circle and a minimum at the highest point. If the velocity of the point mass at the lowest point is large compared with $\sqrt{gl}$, where g is the acceleration due to gravity and l is the length of the rod, then the period (time for one revolution) is approximately $2\pi/\omega$ where ω is the angular velocity (angles in radians) when the string is horizontal (Lamb 1923). In other words a simple rod pendulum is not isochronous when it is rotating continuously.

2.3.5 *Effect of the Earth's Curvature*

The flat Earth assumption implies that gravity acts vertically downwards, and this assumption is used in the derivation of Eq. 2.15 for the period, P, of a *simple rod pendulum.* This is not quite correct for a pendulum on the Earth (Sect. 2.1). An exact solution of the effect of the *Earth's curvature* on the period of a simple rod pendulum requires the use of elliptic integrals (Burko 2003). An approximate solution for a simple rod pendulum at the surface of a sphere, for small amplitudes, is (Rawlings 1993)

$$P = \pi \sqrt{\frac{1}{g\left(\frac{1}{l} + \frac{1}{r}\right)}} \tag{2.26}$$

where P is the pendulum period, g is the acceleration due to gravity, l is pendulum length, and r is the is the radius of the sphere.

As $r \to \infty$ the equation reduces to Eq. 2.15. For finite values of r the pendulum period is reduced. For a 1 m long pendulum on the Earth's surface this reduction is equivalent to about 2.6 s per year (Matthys 2003). Hence, for real pendulums, the effect of the Earth's curvature is so small that, for most practical purposes, it can be neglected.

Mathematically, Eq. 2.26 shows that the pendulum length and the radius of the sphere are interchangeable. The Earth's radius is about 6,000 km. Inserting this and $g = 9.81\ \text{m/s}^2$ into Eq. 2.26 leads to $P \approx 42$ min as $l \to \infty$. This infinitely long pendulum is the *42 minute pendulum.*

2.4 Analysis of a Simple String Pendulum

In a *simple string pendulum* (Fig. 2.1b) the point mass, m, moves on the surface of a sphere. For this reason it is an example of a *spherical pendulum.* Spherical pendulums are sometimes called *conical pendulums* because the pendulum as a whole follows a general conical path. The path taken by the point mass, m, can vary widely, depending on how the simple string pendulum is launched. The behaviour of a simple string pendulum can only be described precisely by relatively simple equations in special cases. In general it displays chaotic behaviour, so behaviour cannot be predicted precisely, although some general features can be deduced.

The inextensible massless string of a simple string pendulum can only resist tensile forces. Hence, for the point mass to be confined to a spherical surface the algebraic sum of the force due to gravity on the point mass plus that due to centripetal force must be tensile. In other words, the string must remain taut. Motions when the string does not remain taut are not considered. In the point mass in sphere analogy (Sect. 2.1) the motion of a simple string pendulum is equivalent to the motion of a point mass, m, moving inside a frictionless sphere under the influence of gravity. The equivalent to the taut string condition is that the normal force between point mass, m, and the inside of the sphere must be compressive. The motions of a simple string pendulum are more easily visualised by using the point mass in sphere analogy. They can be demonstrated by rolling a small, heavy ball, such as ball bearing, in a circular bowl with an approximately spherical region. Figure 2.4 shows a ball bearing, in its rest position, in a turned wooden bowl. The precise path depends upon how the ball bearing is launched.

As shown in Fig. 2.11, the position of a point mass, m, can be specified by using the *spherical polar coordinates*, r, θ, ψ, where r is the distance of the point mass, m, from the centre of the sphere, O, θ is the angle ZOR measured along a *great circle* from the fixed point Z, ψ is the angle between the plane of this great circle and a fixed plane through OZ, and OZ is vertical. A great circle lies on the surface of a sphere with its centre at the centre of the sphere. The angle θ is equivalent to the latitude of a point on the Earth's surface, and the angle ψ to its longitude.

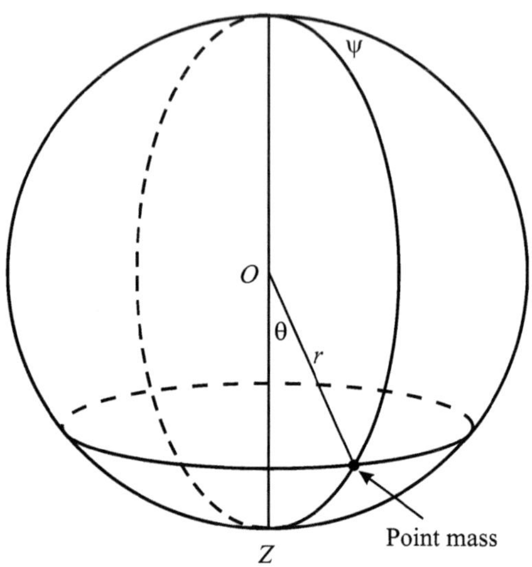

Fig. 2.11 Spherical polar coordinates

If θ changes by a small amount $\delta\theta$, then the point mass, m, moves along an arc $r\delta\theta$ of the great circle through ZP, with velocity $r(\mathrm{d}\theta/\mathrm{d}t)$ where r is the radius of the sphere and of the great circle, and t is time. Similarly, if ψ changes by a small amount $\delta\psi$, then the point mass, m, moves along an arc $r\sin\theta\delta\psi$ of a *small circle* of radius $r\sin\theta$ with velocity $r\sin\theta\ (\mathrm{d}\psi/\mathrm{d}t)$. A small circle is a circle lying on a sphere with radius less than that of the sphere. If the two velocities are combined then the resultant velocity, m_{V}, of the point mass, m, is given by

$$m_{\mathrm{V}} = r\sqrt{\left(\frac{\mathrm{d}\theta}{\mathrm{d}t}\right)^2 + \sin^2\theta\left(\frac{\mathrm{d}\psi}{\mathrm{d}t}\right)^2} \tag{2.27}$$

If the only force on the point mass, m, is that due to gravity, then the equation of energy is (Lamb 1923)

$$\frac{mr^2}{2}\left\{\left(\frac{\mathrm{d}\theta}{\mathrm{d}t}\right)^2 + \sin^2\theta\left(\frac{\mathrm{d}\psi}{\mathrm{d}t}\right)^2\right\} = mgr\cos\theta + \text{const.} \tag{2.28}$$

where r is the radius of the sphere, t is time, and g is the acceleration due to gravity, assumed to be uniform and to act vertically downwards. From the principal of angular momentum

$$\sin^2\theta\frac{\mathrm{d}\psi}{\mathrm{d}t} = h \tag{2.29}$$

where h is a constant. Equations (2.28) and (2.29) lead to

$$\frac{\mathrm{d}^2\theta}{\mathrm{d}t^2} - \frac{h^2\cos\theta}{\sin^2\theta} = -\frac{g}{r}\sin\theta \tag{2.30}$$

It follows that if the path of the point mass, m, passes through the highest point or the lowest point of the sphere, then the path must lie on vertical great circle (Lamb 1923) as shown in Fig. 2.12a. The line joining the point mass, m, to the centre of the sphere rotates about a horizontal axis through the centre of the sphere. Hence the motions of the point mass, m, are analogous to those of a simple rod pendulum (Fig. 2.1a) provided that the normal force between the point mass, m, and the inside of the sphere is compressive. There are two possibilities. In the first the point mass, m, oscillates in the same way as that of a simple rod pendulum (Sects. 2.3 and 2.3.4) provided that the angle θ (Fig. 2.11) is less than 90°. In the second, the point mass, m, rotates about the vertical great circle provided that the angular velocity, ω, at the highest point in the sphere is large enough for there to be a compressive normal force between the point mass, m, and inside of the sphere.

Another possibility arises when $\theta < 90°$ and the path of the point mass, m, is symmetrical about a vertical plane through any point where the direction of motion is horizontal. The path then undulates between two horizontal small circles which it touches alternately at *apses* (Fig. 2.12b). At apses the line joining the point mass, m, to the centre of the sphere rotates about the vertical axis through the centre of the

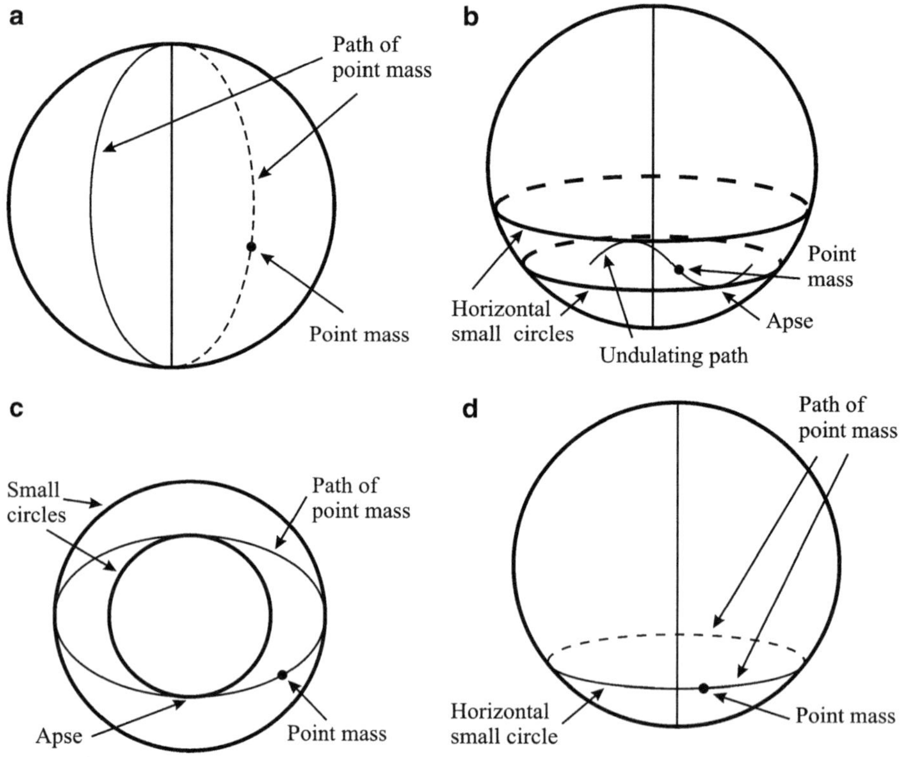

Fig. 2.12 Paths of a point mass moving on the surface of a sphere under the influence of gravity. (**a**) On a vertical great circle. (**b**) Undulating between two horizontal small circles. (**c**) Small amplitude undulating path projected onto a horizontal plane. (**d**) On a horizontal small circle

sphere. At other points on the path the rotation is a combination of rotations about horizontal and vertical axes through the centre of the sphere. The motion of the point mass, m, is periodic, and the value of the angular interval, γ, between adjacent apses is always the same. It is given approximately by (Lamb 1923)

$$\gamma = \frac{\pi}{\sqrt{1 + 3\cos^2\theta}} \tag{2.31}$$

This equation shows that the orbit does not repeat exactly but *precesses* about the vertical axis of the sphere.

For small amplitudes the path is approximately an ellipse (Fig. 2.12c) and the time of swing, T, for one complete traverse of the ellipse is given approximately by

$$T = 2\pi\sqrt{\frac{r}{g}} \tag{2.32}$$

which is Eq. 2.13 with the rod length, l, replaced by r.

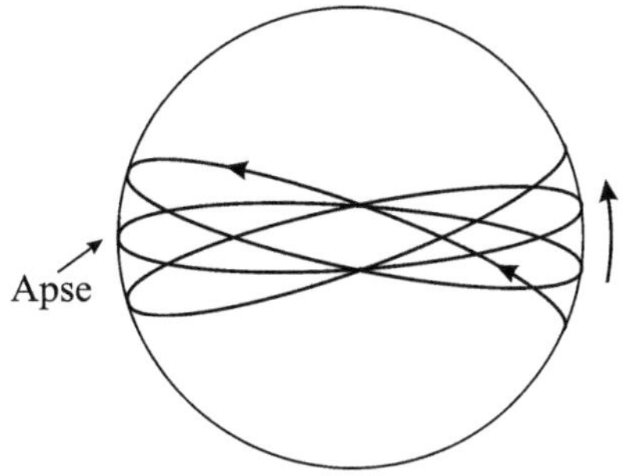

Fig. 2.13 Three complete oscillations of the bob of an ellipsing pendulum. *Arrows* show the directions of the rotation and of the motion of the bob

This is an *elliptical mode of oscillation*, and for small amplitudes a simple string pendulum can be called an *elliptic pendulum*. In the limiting cases the ellipse becomes either a circle or a circular arc. An elliptical mode of oscillation can be resolved, along the major and minor axes of the ellipse, into two planar modes of oscillation along circular arcs. These have the same time of swing, but different amplitudes, and are 90° out of phase. A pendulum swinging in an elliptical mode of oscillation is said to be *ellipsing* (Tobin 2003; Baker and Blackburn 2005; Betrisey 2010). An ellipsing pendulum precesses, as shown schematically in Fig. 2.13 with the rate of precession exaggerated.

If the two small circles shown in Fig. 2.12b coincide then the path of the point mass, m, is a horizontal small circle (Fig. 2.12d). If F is the normal force between the point mass, m, and the inside of the sphere, then the vertical component is $F\cos\theta$, where θ is shown in Fig. 2.11. This is reacted by the force due gravity on the point mass, m, mg, where g is the acceleration due to gravity, assumed to be uniform and acting vertically downwards. Hence

$$F\cos\theta = mg \tag{2.33}$$

From Eq. 2.18 the centripetal force on the point mass m towards the centre of the small circle is $mr_s\,\omega^2$ where r_s is the radius of the small circle ($=r\sin\theta$) and ω is the angular velocity, and is equal to the horizontal component of F, $Fr\sin\theta$. Hence

$$F = mr\omega^2 \tag{2.34}$$

Substituting in Eq. 2.33

$$\omega = \sqrt{\frac{g}{r\cos\theta}} \tag{2.35}$$

and the time of swing, T, for one complete traverse of the small circle is

$$T = \frac{1}{2\pi}\sqrt{\frac{r\cos\theta}{g}} \tag{2.36}$$

In this special case, the analogous simple string pendulum may be called a *circular pendulum*. Replacing r by the pendulum length, l, the time of swing for the

Table 2.3 Circular error, $\sqrt{\cos\theta}$, for a circular pendulum

Pendulum angle θ degrees	$\sqrt{\cos\theta}$	Pendulum angle θ degrees	$\sqrt{\cos\theta}$	Pendulum angle θ degrees	$\sqrt{\cos\theta}$
0	1.000000	4	0.998781	30	0.930605
1	0.999924	5	0.998010	45	0.840896
2	0.999695	10	0.992375	60	0.707107
3	0.999315	20	0.969377	90	0

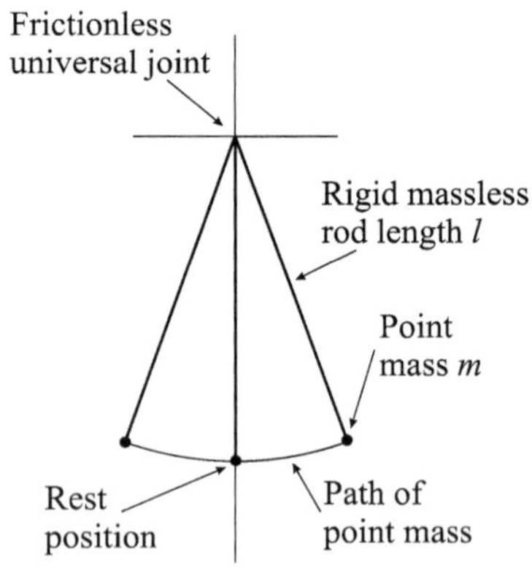

Fig. 2.14 Spherical rod pendulum

analogous circular pendulum is given by

$$T = \frac{1}{2\pi}\sqrt{\frac{l\cos\theta}{g}} \tag{2.37}$$

and the circular error is $\sqrt{\cos\theta}$. Some values of $\sqrt{\cos\theta}$ are shown in Table 2.3. The time of swing of a circular pendulum decreases as the amplitude increases, whereas it increases for a simple rod pendulum (Table 2.2).

2.5 Spherical Rod Pendulum

The *spherical rod pendulum* (Fig. 2.14) is a simple rod pendulum (Fig. 2.1a) with the frictionless pivot replaced by a frictionless universal joint (Barnard 1874). Its point mass moves on a sphere so it is a *spherical pendulum*. A spherical rod pendulum has two degrees of freedom and it can display chaotic behaviour. All the motions of the simple rod pendulum (Sect. 2.3) and the simple string pendulum (previous section) are possible, together with some additional motions. For small amplitudes its motions are the same as those of a simple string pendulum (previous section).

References

Baker GL, Blackburn JA (2005) The pendulum. A case study in physics. Oxford University Press, Oxford

Barnard JG (1874) Problems of rotary motion presented by the gyroscope, the precession of the equinoxes, and the pendulum. Smithsonian Contributions to Knowledge, 19(240):1–52

Bateman D (1977) Vibration theory and clocks. Part 1. Harmonic motion. Horological J 120(1):3–8

Berléndaz A, Francés J, Ortuño M, Gallego S, Bernabeu JG (2010) Highly accurate approximation solutions for the simple pendulum in terms of elementary functions. Eur J Phys 31(3):L65-L70

Betrisey M (2010) Works of art that also tell the time. http.//www.betrisey.ch/eindex.htm Accessed 20 July 2010

Britten FJ (1978) The watch & clock makers' handbook, dictionary and guide, 16th edn. Arco Publishing Company, New York (Revised by Good R)

Burko L (2003) Effects of the spherical earth on a simple pendulum. Eur J Phys 24(2):125–130

Camponario JM (2006) Using textbook errors to teach physics: examples of specific activities. Eur J Phys 27(4):975–981

Feinstein GF (1995) M Fedchenko and his pendulum astronomical clocks. NAWCC Bull 1995:169–184

Huygens C (1673/1986) Horologium oscillatorium. The pendulum clock or geometrical demonstrations concerning the motion of pendula as applied to clocks. The Iowa State University Press, Iowa. (Trans by Blackwell, Ames RJ)

Johannessen K (2010) An approximate solution to the equation of motion for large-angle oscillations of the simple pendulum with initial velocity. Eur J Phys 31(3):511–518

Kreyszig E (1983) Advanced engineering mathematics, 5th edn. Wiley, New York

Lamb H (1923) Dynamics, 2nd edn. Cambridge University Press, Cambridge

Loney SL (1913) An elementary treatise on the dynamics of a particle and of rigid bodies. Cambridge University Press, Cambridge

Matthews RJ (2000) Time for science education. How teaching the history and philosophy of pendulum motion can contribute to science literacy. Kluwer Academic/Plenum Publishers, New York

Matthys R (2003) The period of a simple pendulum is not $2\pi\sqrt{L/g}$. Horological Sci Newsl (2003-5), 13

Parwani RR (2004) An approximate expression for the large angle period of a simple pendulum. Eur J Phys 25(1):37–39

Rawlings AL (1993) The science of clocks and watches, 3rd edn. British Horological Institute, Upton

Tobin W (2003) The life and science of Léon Foucault. The man who proved that the earth rotates. Cambridge University Press, Cambridge

Whelan PM, Hodgson MJ (1977) Essential pre-university physics. John Murray, London

Chapter 3
Variations on Simple Pendulums

3.1 Introduction

The simple rod pendulum (Sect. 2.3) is a convenient idealisation of the real pendulums often used to control clocks. *Compound pendulum* is the term used by horologists for a real clock pendulum, as opposed to an simple rod pendulum where the rod is assumed to be rigid and massless, with pendulum mass concentrated at a point. In a real clock pendulum, the rod (Figs. 1.1 and 1.2b) is relatively light, but not rigid and massless, and the bob is heavy, but its mass is not concentrated at a point. The term is also applied to an idealisation of a compound pendulum that are used in order to make mathematical analysis more tractable (Lamb 1923; Rawlings 1993). This idealisation is discussed in this chapter.

Analysis of simple rod pendulums and idealised compound pendulums does not always provide an explanation of the behaviour of real clock pendulums, so other idealisations are sometimes needed. Other types of pendulum are used in scientific, industrial and recreational applications, and appropriate idealisations are needed. Idealisations of several types of pendulum are described in this chapter. Except for idealisations of compound pendulums, these idealisations are all assembled from components of simple rod pendulums or simple string pendulums (Sect. 2.4). Most can display chaotic behaviour.

3.2 Compound Pendulum

An idealised *compound pendulum* is a rigid body suspended from a frictionless horizontal pivot. Rigid means that the body does not deform under an applied load. It is assumed that the only forces acting are those due to gravity, acting vertically downwards, on the component parts of the body. A clock pendulum as an idealised compound pendulum is shown in Fig. 3.1.

L.P. Pook, *Understanding Pendulums: A Brief Introduction*, History of Mechanism and Machine Science 12, DOI 10.1007/978-94-007-1415-1_3,

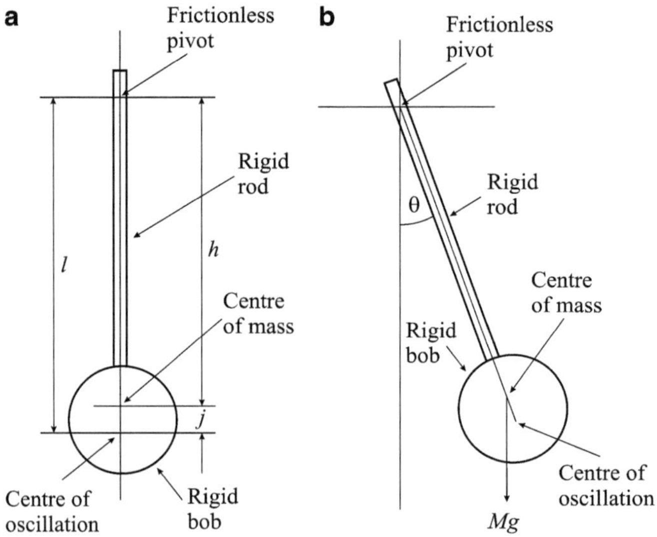

Fig. 3.1 A clock pendulum as an idealised compound pendulum. (**a**) Rest position. (**b**) Deflected

For a body that can rotate about an axis, if r is the perpendicular distance from the axis of an element mass m, then the summation $\sum mr^2$ is the *moment of inertia*, I, of the body about the axis. For simple shapes values of I can be obtained by integration and are listed in textbooks, for example Loney (1913). If M is the mass of the body and taking

$$M\kappa^2 = \sum mr^2 \tag{3.1}$$

then κ is the *radius of gyration* of the body about the axis. For a simple rod pendulum (Fig. 2.1a) the moment of inertia of a is ml^2 where m is the point mass and l is the length of the rod, and the radius of gyration about the axis (frictionless pivot) is l.

For small amplitudes (value of θ in Fig. 3.1b) the derivation of the pendulum period, P, follows that for the simple rod pendulum given in Sect. 2.3 (Loney 1913; Lamb 1923). The moment about the frictionless pivot is $Mgh\sin\theta$ where g is the acceleration due to gravity, h is the distance of the *centre of mass*, often called the *centre of gravity*, from the frictionless pivot, and θ is the pendulum angle. Hence

$$I\frac{\mathrm{d}^2\theta}{\mathrm{d}t^2} = -Mgh\sin\theta \tag{3.2}$$

where I is the moment of inertia, and t is time. This is the same as for a simple rod pendulum if

$$l = I/Mh = \kappa^2/h \tag{3.3}$$

where l is the *effective length* of a compound pendulum, and

$$l = \kappa^2/h \tag{3.4}$$

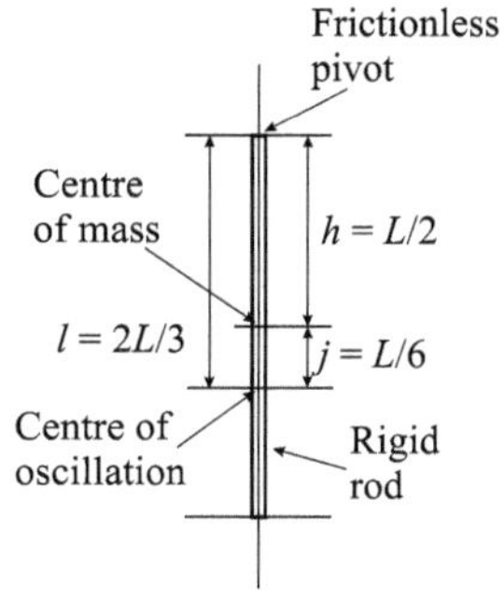

Fig. 3.2 Rod pendulum, length L

Hence, the pendulum period, P, is given by Eq. 2.15, with the length of a simple rod pendulum replaced by the effective length of an idealised compound pendulum. Further, the circular error of an idealised compound pendulum is the same as that of a simple rod pendulum (Sect. 2.3.4).

A point known as the *centre of oscillation* (Saunier 1861/1975; Britten 1892) lies on a straight line through the frictionless pivot and the centre of mass at distance l from the frictionless pivot and j from the centre of mass. It can be shown that (Lamb 1923)

$$\kappa_{\mathrm{m}}^2 = \kappa^2 - h^2 = hj \tag{3.5}$$

where κ_{m} is the radius of gyration about the centre of mass and j is the distance from the centre of mass to the centre of oscillation. This implies, as first pointed out by Huygens (1673/1986), that the frictionless pivot and the centre of oscillation are interchangeable. That is, if the pendulum were pivoted at the centre of oscillation, then the previous frictionless pivot would become the new centre of oscillation, and the pendulum period, P, would be the same. The centre of oscillation is sometimes called the *centre of percussion* (Rawlings 1993). This is because if a moving idealised compound pendulum is stopped at the centre of percussion there is no reaction at the frictionless pivot.

As an example, consider an idealised compound pendulum consisting of a rigid uniform rod, length L, with a frictionless pivot at one end (Fig. 3.2). The moment of inertia, I, is $ML^2/3$, where M is the mass of the rod, the radius of gyration, κ, is $L/\sqrt{3}$, and the distance of the centre of mass from the frictionless pivot, h, is $L/2$. Hence, from Eq. 3.4, the effective length, l, of the compound pendulum is $2L/3$.

3.3 Double Rod Pendulum

A *double rod pendulum* consists of two simple rod pendulums (Fig. 2.1a) arranged in series, as shown in Fig. 3.3 (Lamb 1923; Kibble and Berkshire 1996; Baker and Blackburn 2005; Vankó 2007). The upper rigid massless rod, length l_1, is suspended at its upper end from a horizontal frictionless pivot, and has a point mass, m_1, at its

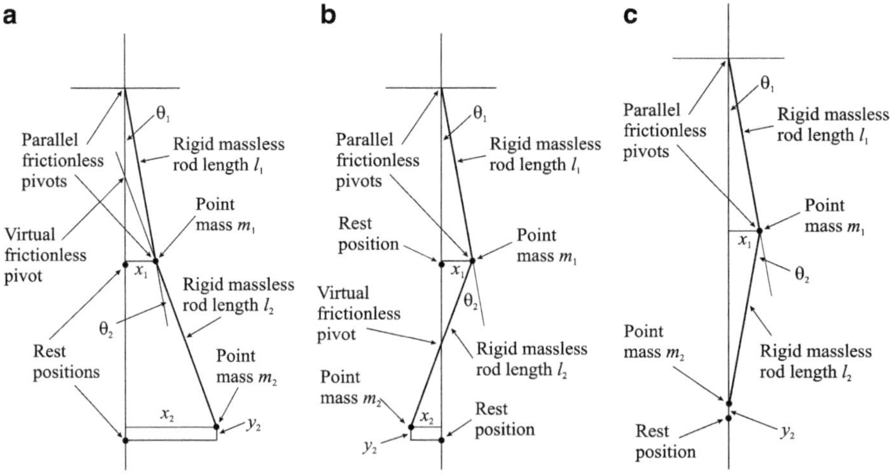

Fig. 3.3 Double rod pendulum. (**a**) Pendulum angles and point masses moving in phase. (**b**) Pendulum angles and point masses moving out of phase. (**c**) Pendulum angles moving out of phase. Point mass m_2 moving vertically

lower end. The lower rigid massless rod, length l_2, is suspended at its upper end from a horizontal frictionless pivot at the lower end of the upper rod, and has a point mass, m_2, at its lower end. The two frictionless pivots are parallel so the motion of the point masses is confined to a vertical plane, and there are two degrees of freedom. The pendulum angle of the upper rod from the vertical is θ_1, and that of the lower rod from the upper rod is θ_2. The horizontal components of displacements of the point masses m_1 and m_2 from their rest positions are x_1 and x_2 respectively. Four parameters, l_1, l_2, m_1 and m_2 are needed to define a double rod pendulum, so there is a wide range of possibilities, and there is no such thing as a typical double rod pendulum. A Meccano model of a double rod pendulum is shown in Fig. 3.4. In the model, the pendulums are arranged so that they can swing in complete circles. In general, there are no simple expressions for times of swing. The governing equations are complex and, except for some special cases, can only be solved numerically (Kibble and Berkshire 1996).

For small amplitudes a double rod pendulum has two modes of oscillation (Lamb 1923). The frictionless pivots are parallel so these are planar modes of oscillation. In both modes the lower rod oscillates about a virtual frictionless pivot. Which mode appears depends on how the pendulum is launched. Both modes of oscillation can be demonstrated by using the Meccano model. In the first mode of oscillation both the pendulum angles, θ_1 and θ_2, and the point masses m_1 and m_2 move *in phase*, that is the *phase angle*, ϕ, of both point masses at a given time is the same (Fig. 3.3a). Hence the *phase*, ε. which is the difference between the phase angles is zero. The horizontal components of the displacements of m_1 and m_2, x_1 and x_2, have the same sign. In the second mode of oscillation the pendulum angles, θ_1 and θ_2 move *out*

Fig. 3.4 Meccano model of a double rod pendulum

of phase, that is the phase angle is 180°. For a given double rod pendulum the frequency of oscillation is the second mode of oscillation is always greater than in the first mode of oscillation.

Geometric considerations show that, for the second mode of oscillation there are two possibilities for the point masses. In the first (Fig. 3.3b) they, and the horizontal displacements, x_1 and x_2, move out of phase so have opposite signs. The point mass m_2 moves up and down at twice the pendulum frequency. The maximum values of the vertical displacement, y_2, are at the ends of the arc followed by point mass m_1. In the special case shown in Fig. 3.3c x_2 is zero. The position of the virtual frictionless pivot of the lower rod is indeterminate, and behaviour can be chaotic.

Lamb (1923) gives solutions for some special cases. For $m_1 \ll m_2$, in the first mode of oscillation $\theta_2 \approx$ zero and the pendulum oscillates much like a simple rod pendulum of length $(l_1 + l_2)$ The time of swing, T, is given approximately by (cf Eq. 2.13)

$$T = 2\pi\sqrt{\frac{l_1 + l_2}{g}} \tag{3.6}$$

In the second mode of oscillation $x_1 \gg x_2$ and $y_2 \approx x_2$ that is the lower point mass, m_2, is nearly stationary (Fig. 3.3c). For $l_1 \approx l_2$ the upper point mass, m_1, oscillates rather like a point mass m_1, attached to a massless elastic string under a tension

m_2g, stretched between fixed points (Fig. 1.13). This is the *point mass on a string analogy*, and the time of swing, T, is given approximately by

$$T = \pi \sqrt{\frac{m_1 (l_1 + l_2)}{m_2 g}} \tag{3.7}$$

which is much shorter than that given by Eq. 3.6.

For $m_1 \gg m_2$ and $l_1 \approx l_2$ a double rod pendulum displays chaotic behaviour, as noted by Lamb (1923), who does not use that term. The double rod pendulum is one of the simplest mechanical systems that can display chaotic behaviour. In particular, a plot of the angular velocity of the lower rigid massless rod against time is chaotic (Baker and Blackburn 2005). Whether or not chaotic behaviour appears depends on how the pendulum is launched. However, chaotic behaviour becomes increasingly likely as the amplitude increases. There are videos of chaotic behaviour, on YouTube, of both physical models of double pendulums and computer simulations (Anonymous 2010). Chaotic behaviour can be demonstrated by using the Meccano model. The appeal of a chaotic mechanical system is essentially visual, and the double rod pendulum is also an amusing toy.

In general, when the Meccano model is launched, its motions are chaotic, with no apparent relationship between the motions of the bobs. However, as the motions decay, due to the effects of friction etc., the motions tend to degenerate into the first mode of oscillation (Fig. 3.3a). The two parts of a double rod pendulum can be regarded as *coupled pendulums* in which energy is transferred between pendulums. Coupled pendulums have a tendency to become synchronised, as first observed by Huygens in 1665 (Baker and Blackburn 2005), and recently discussed by Haine (2010).

3.4 Blackburn Pendulum

The discovery of the *Blackburn pendulum*, also called *Blackburn's pendulum*, is sometimes attributed to Hugh Blackburn, who first described it in 1844 (Benham 1909; Ashton 2003). However, it was discussed by James Dean in 1815 and analysed mathematically by Nathaniel Bowditch later in the same year (Whitaker 1991, 2005). It appears to have been re-discovered independently by Blackburn. A Blackburn pendulum is a string pendulum, (Lamb 1923; Whitaker 1991, 2005; Baker and Blackburn 2005) with three inextensible, massless strings arranged in a Y shape (Fig. 3.5). The upper ends of the Y are clamped, and a point mass. m, is attached to the lower end. The distance from a line through the clamps to the centre of the Y is l_1, and the distance from the centre of the Y to the point mass, m, is l_2. It is assumed that the forces on the point mass are such that the forces in all three strings are tensile. Otherwise, a Blackburn pendulum does not maintain the configuration

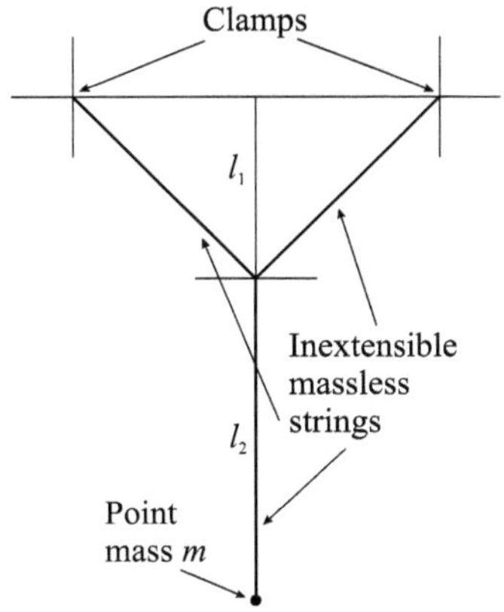

Fig. 3.5 Blackburn pendulum

described above. A different type of pendulum, invented by J A Blackburn (Baker and Gollub 1996), and used for studying chaotic behaviour, is also known as the Blackburn pendulum.

The tension in the strings due to forces on the point mass, m, means that the Y is always flat. Two *orthogonal* (at right angles) modes of oscillation are possible, so it has two degrees of freedom. In the first mode of oscillation the Y oscillates as a whole about the clamps and, for small amplitudes, the pendulum is equivalent to a simple rod pendulum (Fig. 2.1a) length $l_1 + l_2$ so the approximate time of swing, T, is given by Eq. 3.6 where g is the acceleration due to gravity. In the second mode the lower string oscillates about the centre of the Y in the plane of the Y (the plane of the paper in Fig. 3.5) and the approximate time of swing is given by

$$T = 2\pi\sqrt{\frac{l_2}{g}} \tag{3.8}$$

More generally, combinations of the two modes of oscillation are possible and the point mass, m_2, moves on the surface of a torus generated by the rotation of a circle with centre at the centre of the Y and radius l_2 about the line joining the clamps.

3.5 Bifilar Pendulum

In a *bifilar pendulum* (Gauld 2005) a point mass, m, is suspended from a pair of inextensible massless strings of equal length arranged so that the path of the point mass, m, is constrained to a vertical plane. A side view of a bifilar pendulum in the rest position shown in Fig. 3.6. The behaviour of a bifilar pendulum is the same as that of a simple rod pendulum (Sect. 2.3), provided that the strings remain taut. Hence, if the resolved length of the strings in a vertical direction is l, then for small amplitudes the pendulum period, P, given approximately by Eq. 2.15.

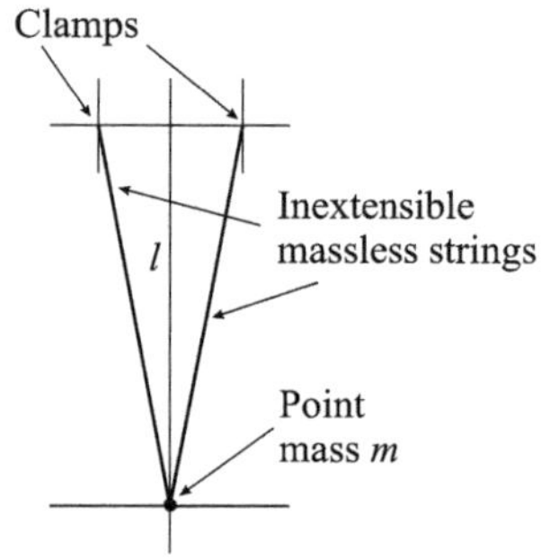

Fig. 3.6 Side view of a bifilar pendulum in the rest position

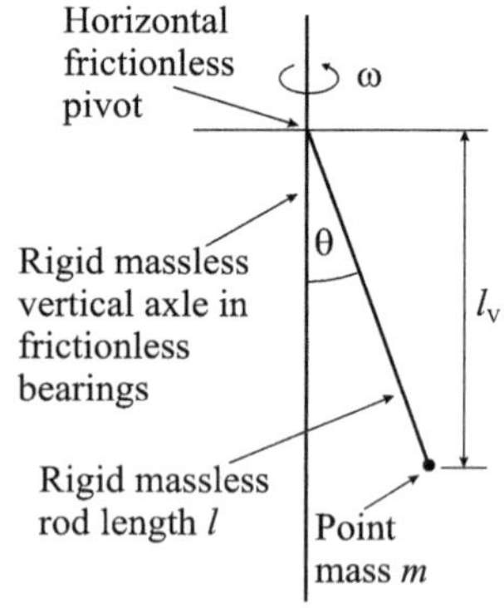

Fig. 3.7 Rotating simple rod pendulum. If the angular velocity, ω, is constant it is a circular rod pendulum

3.6 Rotating Simple Rod Pendulum

The *rotating simple rod pendulum* (Fig. 3.7), is a simple rod pendulum (Fig. 2.1a) with the frictionless pivot on a rigid massless vertical axle free to rotate in frictionless bearings. The point mass, m, moves on the surface of a sphere so it is a *spherical pendulum.* Kibble and Berkshire (1996) call it a *spherical rod pendulum.* In general, the motions of its point mass, m, are the same as that of the simple string pendulum (Sect. 2.4). However, motions are restricted because the point mass, m, cannot pass through the axle.

If the pendulum rotes at constant angular velocity, ω, the point mass, m, moves on a horizontal small circle (Fig. 2.12d). In this special case a rotating simple rod pendulum can be called a *circular pendulum,* or a *conical pendulum.* The time of swing for one complete revolution of the small circle is given by Eq. 2.37. From this equation (Lineham 1914), the vertical distance of the point mass, m, below the pivot, l_V (Fig. 3.7) is

$$l_V = \frac{g}{\omega^2} \tag{3.9}$$

where g is the acceleration due to gravity, and ω is the angular velocity.

3.7 Quadrifilar Pendulum

A *quadrifilar pendulum* is shown in Fig. 3.8. It consists of a uniform rigid square plate, diagonal, $2c$, mass, M, suspended by four vertical inextensible massless strings, length, l. In effect, it is four coupled simple string pendulums (Sect. 2.4). Provided that the strings remain taut a quadrifilar pendulum has three degrees of freedom and, because the strings are of equal length, the plate remains horizontal. If the rigid plate does not rotate then there are two degrees of freedom, and the motion of each point mass, m, is in phase and identical and also, in general, identical to the motion of the point mass, m, of a simple string pendulum (Sect. 2.4). Exceptions are due to inference between components of the quadrifilar pendulum. For small amplitudes the time of swing, T, is given by Eq. 2.13 where g is the acceleration due to gravity.

The third degree of freedom means that a *torsional mode of oscillation* is possible. In this torsional mode of oscillation the uniform rigid square plate rotates about a vertical axis through its centre. If the plate is turned through a small angle, θ, about its centre (Fig. 3.8c) then the lower end of each string moves through an arc of length approximately $c\theta$, where $2c$ is a diagonal of the square and θ is in radians. The tensile force in each string is approximately $^1/_4 Mg$, where M is the mass of the plate, and g is the acceleration due to gravity. The horizontal component of the tensile force is approximately $^1/_4 Mgc0/l$, and the total restoring couple is $Mgc^2\theta/l$. The corresponding differential equation is (Lamb 1923)

$$M\kappa^2\frac{d^2\theta}{dt^2} = -Mg\frac{c^2}{l}\theta \tag{3.10}$$

where κ is the radius of gyration of the square, and t is time. The approximate time of swing, T, is

$$T = 2\pi\sqrt{\frac{\kappa^2 l}{gc^2}} \tag{3.11}$$

which differs from Eq. 2.13 by the factor κ/c.

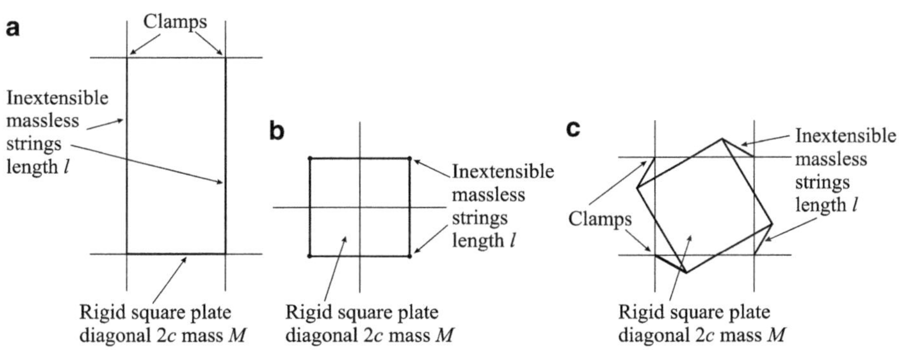

Fig. 3.8 Quadrifilar pendulum. (**a**) Side view. (**b**) Top view. (**c**) Top view, plate rotated through angle θ

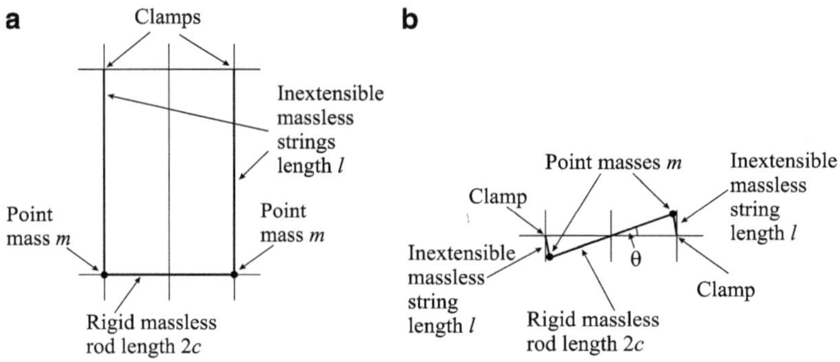

Fig. 3.9 Dual string pendulum. (**a**) In rest position, view from front. (**b**) Rigid massless rod rotated through angle θ, view from top

3.7.1 Dual String Pendulum

A *dual string pendulum* is a simplification of the quadrifilar pendulum (previous section), and its motions are analogous. It consists of two identical simple string pendulums (Sect. 2.4) with the point masses connected by a rigid massless rod, length $2c$. In the rest position the strings are vertical and the rigid massless rod is horizontal, as shown in Fig. 3.9a. Provided that the strings remain taut a dual string pendulum has three degrees of freedom. If the rigid massless rod remains parallel to its rest position then there are two degrees of freedom, and the motion of each point mass, m, is in phase and identical and also, in general, identical to the motion of the point mass, m, of a simple string pendulum (Sect. 2.4). Exceptions are due to inference between components of the dual string pendulum.

The third degree of freedom means that a torsional mode of oscillation is possible, in which the rigid massless rod rotates about a vertical axis (Fig.3.9b). In this torsional mode of oscillation the simple string pendulums are moving out of phase, that is the phase is 180°. For the dual string pendulum the radius of gyration, κ, of the rod and point masses about the vertical axis through its centre is c, so Eq. 3.11 reduces to Eq. 2.13. Hence, for small amplitudes the time of swing is independent of the length of the rigid massless rod.

3.8 Trapezium Pendulum

In mechanical engineering terms the *trapezium pendulum* is a *Watt's linkage*, developed in 1784 (Trébaol 2008), and sometimes called *Watt's parallel motion* (Dunkerley 1910). Watt's linkage is an example of a *four bar linkage*. In a four bar linkage the four links are pivoted together so that they move within a plane, as shown schematically in Fig. 3.10, These links are: a static bar, regarded as being fixed in

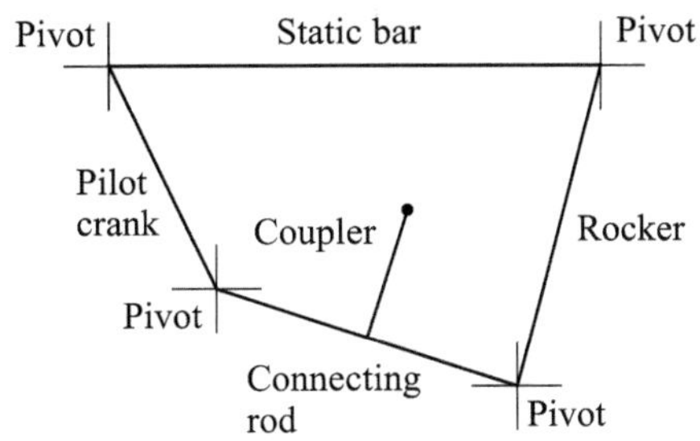

Fig. 3.10 Four bar linkage

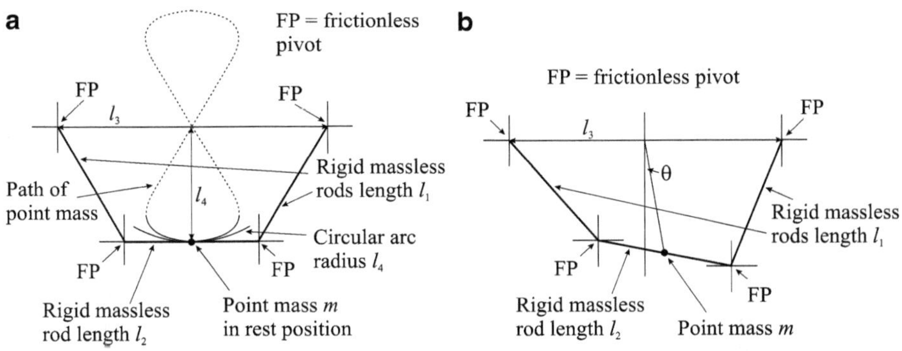

Fig. 3.11 Trapezium pendulum. (**a**) In rest position. (**b**) Definition of pendulum angle θ

space, a pilot crank, a connecting rod, and a rocker. A coupler may be attached to one of the moving links. A four bar linkage has one degree of freedom since one parameter, for example the angle between the static bar and the pilot crank, is needed to specify its configuration.

The trapezium pendulum as a form of Watt's linkage is shown in Fig. 3.11. The pilot crank and rocker of a four bar linkage are the two rigid massless rods, length l_1, shown in the figure. At their lower ends they are connected by horizontal frictionless pivots to a rigid massless rod, length l_2, with a point mass, m, at its centre. This is horizontal in the rest position (Fig. 3.11a), and is the connecting rod in a four bar linkage. At their upper ends the two rigid massless rods, length l_1, are fixed in space by two fixed horizontal frictionless pivots, at the same level, and a distance l_3 apart. The two fixed frictionless pivots correspond to the static bar in a four bar linkage. The complete path of the point mass, m, is a figure of eight with two nearly straight portions, which are exploited in Watt's parallel motion. Part of the path, shown by a solid line in Fig. 3.11a, is a non circular arc.

Design methods for four bar linkages in general (Hrones and Nelson 1951), and Watt's linkage in particular, are well established (Dunkerley 1910; Trébaol 2008). However, equations for the motions of four bar linkages are cumbersome, and there is no simple expression for the motion of the point mass, m, in the trapezium pendulum, Hence there is, in general, no simple expression for its time of swing.

At the rest position the path of the point mass, m, has radius l_4, where l_4 is the vertical distance from the fixed frictionless pivots to the point mass (Fig. 3.11a). Hence, from Eq. 2.13 for a simple rod pendulum, the time of swing, T, for small amplitudes is given approximately by

$$T = 2\pi\sqrt{\frac{l_4}{g}} \tag{3.12}$$

where g is the acceleration due to gravity. The path that would be followed by the point mass, m, of a simple rod pendulum, length l_4. is shown by the circular arc in Fig. 3.11a. For the same pendulum angle, θ, (Figs. 2.2 and 3.11b) the path of the point mass, m, of the trapezium pendulum deviates increasingly from the circular arc as the pendulum angle, θ, increases. In consequence, the time of swing, T, of the trapezium pendulum decreases relative to that of a simple rod pendulum.

3.8.1 Dual Rod Pendulum

In the special case where $l_2 = l_3$ the pilot crank and rocker of a four bar linkage are parallel, and a trapezium pendulum becomes a *dual rod pendulum.* A dual rod pendulum consists of two rigid massless rods, length l, suspended from frictionless pivots connected to a horizontal rigid massless rod with a point mass, m, at its centre. In the rest position the rigid massless rods, length l, are vertical, as shown in Fig. 3.12a. The motion of the point mass, m, is identical to that of a simple rod pendulum (Sect. 2.3) except that, because of interference between the links, the pendulum angle, θ, (Fig. 3.12b) cannot exceed 90°. For small amplitudes the time of swing, T, is given approximately by Eq. 2.13 where g is the acceleration due to gravity.

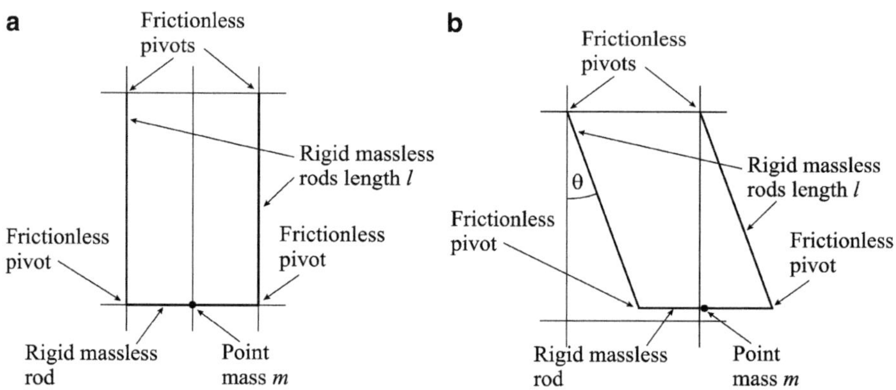

Fig. 3.12 Dual rod pendulum. (**a**) In rest position. (**b**) Deflected

3.9 Double String Pendulum

A *double string pendulum* (Fig. 3.13) consists of two simple string pendulums (Fig. 1.4b) arranged in series. The upper inextensible massless string, length l_1, is clamped at its upper end, and has a point mass, m_1, at its lower end. The lower inextensible massless string, length l_2, is attached at its upper end to the point mass m_1, and has a point mass, m_2, at its lower end. A double string pendulum has four degrees of freedom so possible motions are extremely complicated, and it can display chaotic behaviour. If the point masses m_1 and m_2 remain in a vertical plane, the planar modes of oscillation are analogous to the modes of oscillation of a double rod pendulum (Sect. 3.3) provided that the strings remain taut. Modes of oscillation in which the point masses move in ellipses are possible so a double string pendulum can be called a *twin-elliptic pendulum*.

A *twin-elliptic harmonograph* (Whitaker 2001), is a twin-elliptic pendulum that is used for entertainment. It can be regarded as a real version of a double string pendulum. A Meccano model of a twin-elliptic harmonograph is shown in Fig. 3.14. This is based on a design by Anonymous (1955). The upper pendulum rod of a twin–elliptic harmonograph is suspended by a gimbal and, together with its bob, is sometimes called the *main pendulum*, and the lower pendulum rod, suspended by a short string from the main pendulum, is sometimes called the *deflector pendulum*. A twin-elliptic harmonograph includes provision for recording the motion of an upward extension of the upper rod. Some examples are given in Sect. 8.5.3.

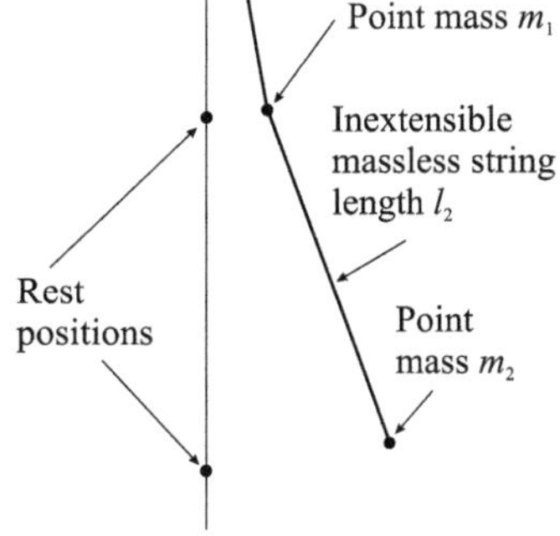

Fig. 3.13 Double string pendulum

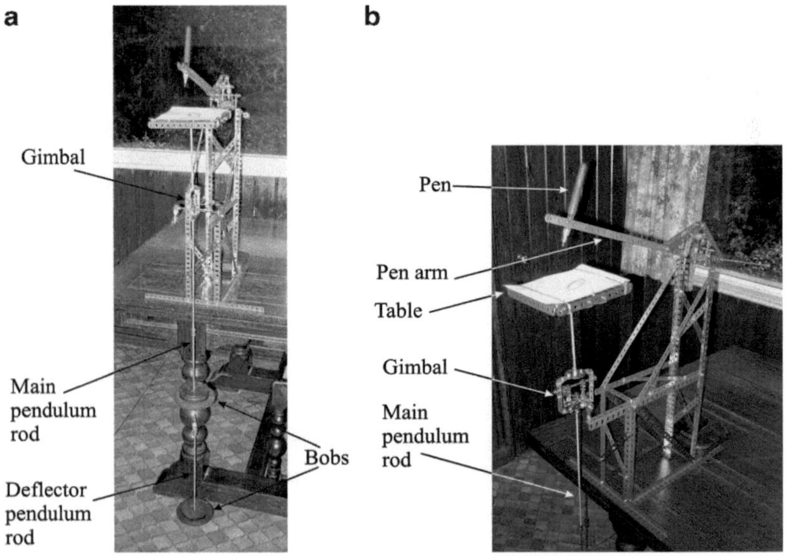

Fig. 3.14 Meccano twin-elliptic harmonograph

References

Anonymous (1955) A fascinating designing machine. Meccano Magazine, 40(7), 384–385

Anonymous (2010) Double pendulum animation. http://www.youtube.com/results?search_query=double+pendulum+animation&search_type=&aq=3&oq=double+pendulum

Ashton A (2003) Harmonograph. A. visual guide to the mathematics of music Walker & Company, New York

Baker GL, Blackburn JA (2005) The pendulum. A case study in physics. Oxford University Press, Oxford

Baker GL, Gollub JP (1996) Chaotic dynamics. An introduction, 2nd edn. Cambridge University Press, Cambridge

Benham CE (1909) Descriptive and practical details as to harmonographs. In: Newton HC (ed) Harmonic vibrations and vibration figures. Newton and Co., London, pp 26–86

Britten FJ (1892) The watch & clockmakers' handbook, dictionary and guide, 8th edn. E & FN Spon, London

Dunkerley S (1910) Mechanisms, 3rd edn. Longmans, Green and Co., London

Gauld C (2005) Pendulums in the physics education literature: a bibliography. In: Matthews MR, Gauld CF, Stinner A (eds) The pendulum. Scientific, historical and educational perspectives. Springer, Dordrecht, pp 505–526

Haine J (2010) Synchronous oscillations of asymmetric coupled pendulums. Horological Sci Newsl (2010–5): 6–13

Hrones JA, Nelson GL (1951) Analysis of the four-bar linkage. Its application to the synthesis of mechanisms. Technology Press of the Massachusetts Institute of Technology. Wiley, New York

Huygens C (1673/1986) Horologium oscillatorium. The pendulum clock or geometrical demonstrations concerning the motion of pendula as applied to clocks. Translated by Blackwell RJ. The Iowa State University Press, Ames

Kibble TWB, Berkshire FH (1996) Classical mechanics, 4th edn. Longman, Harlow

Lamb H (1923) Dynamics, 2nd edn. Cambridge University Press, Cambridge

Lineham WJ (1914) A textbook of mechanical engineering, 11th edn. Chapman and Hall, London

Loney SL (1913) An elementary treatise on the dynamics of a particle and of rigid bodies. Cambridge University Press, Cambridge

Rawlings AL (1993) The science of clocks and watches, 3rd edn. British Horological Institute Ltd., Upton

Saunier C (1861/1975) Treatise on modern horology. W & G Foyle Ltd, London. (Trans by Tripplin J, Rigg E)

Trébaol G (2008) Simulation of four bar linkages. http://gtrebaol.free.fr/doc/flash/four_bar/doc/ Accessed 8 Feb 2010

Vankó P (2007) Investigation of a chaotic double pendulum in the Basic Physics Level Teaching Laboratory. Eur J Phys 28(1):61–69

Whitaker RJ (1991) A note on the Blackburn pendulum. Am J Phy 59(4):330–333

Whitaker RJ (2001) Harmonographs I. Pendulum design. Am J Phy 69(2):162–173

Whitaker RJ (2005) Types of two-dimensional pendulums and their uses in education. In: Matthews MR, Gauld CF, Stinner A (eds) The pendulum. Scientific, historical and educational perspectives. Springer, Dordrecht, pp 377–391

Chapter 4
Pendulum Clocks

4.1 Introduction

In modern usage '*clock*' refers to any device for measuring and displaying time, but not intended for carrying on the person. Clocks which do not strike or chime are sometimes, pedantically, called *timepieces* (Britten 1978). A *pendulum clock* has five essential features (Anonymous 2010). Firstly, a timekeeping element: this is its pendulum which oscillates at a constant frequency. Secondly, a power source, usually either a falling weight, or a coiled spring called a *mainspring*: this replaces energy lost due to inevitable friction etc. Thirdly, a *controlling device* called the *escapement*: this has the dual function of providing *impulses* to replace energy lost due to friction etc., and converting the pendulum oscillations into a series of pulses to measure time. The escapement is usually connected to the pendulum by a crank called a *crutch* (Figs. 1.2a and 4.1b). The crutch has a fork at its lower end that engages the pendulum. In an escapement, one tooth of the *escape wheel* (usually called the '*scape wheel*) is released at regular intervals. Fourthly, a *counter train*, this is the *going train* of gears: this counts and sums the pulses into hours and minutes etc. The going train usually includes a friction device allowing the time to be set manually. Finally, an *indicating device*, usually a dial with hands that indicate hours, minutes and seconds. The provision of appropriately timed impulses to an oscillating pendulum in order to replace lost energy, or to increase its amplitude, is sometimes called *pumping*, but this term is not used in a horological context.

Huygens is usually credited with the invention of the pendulum clock in the seventeenth Century, but he did not claim to be the inventor (Edwardes 1981), and there is evidence that he was not the inventor (Edwardes 1981; Penney 2009; King 2009). What probably happened is that the pendulum clock was invented independently by more than one person. The invention resulted in a dramatic improvement in timekeeping. Obviously, the accuracy of a pendulum clock depends upon the accuracy with which its pendulum keeps time. Essential features of a pendulum clock are that the pendulum oscillation must be stable in the presence of small disturbances, and that just enough energy must be supplied to compensate for

L.P. Pook, *Understanding Pendulums: A Brief Introduction*, History of Mechanism and Machine Science 12, DOI 10.1007/978-94-007-1415-1_4,

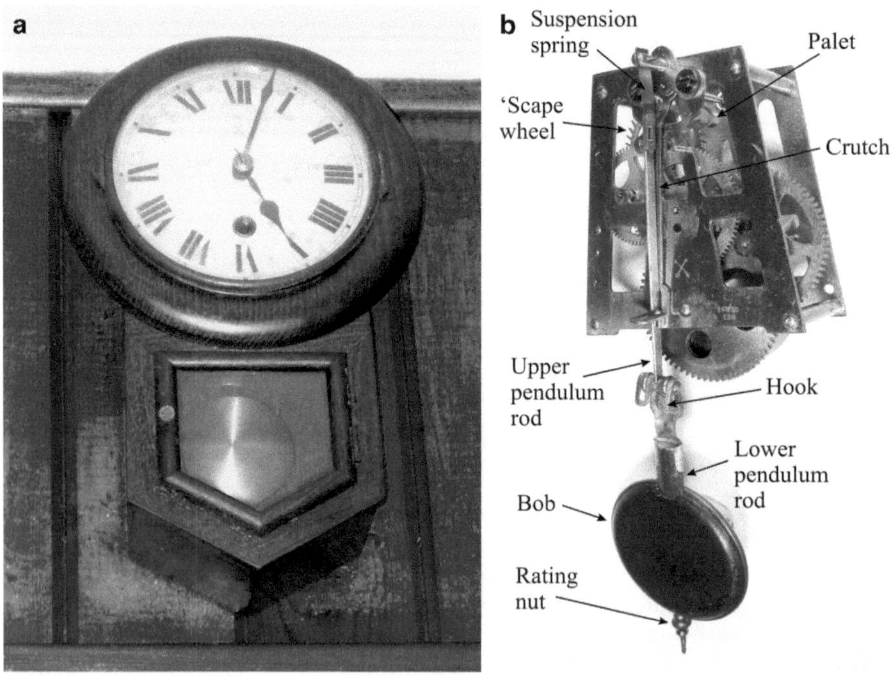

Fig. 4.1 Early twentieth century wall clock. (**a**) General view. (**b**) Rear view of movement. Pendulum length $17^1/_2$ cm from top of suspension spring to centre of bob

the inevitable losses due to friction etc. In mechanical engineering terms a pendulum clock is a *dynamical system* that must be considered as a whole (Denny 2002; Heldman 2009).

What Huygens did do was to provide the first theoretical and practical account of the application of pendulums to clocks (Huygens 1658). In the three and a half centuries since then the history of the pendulum clock is very much the history of refinements to improve timekeeping (Edwardes 1977; Rawlings 1993; Roberts 2003, 2004). Refinements, and attempted refinements, are so numerous that it is impossible to write a concise history. As often happens in technological developments, some refinements were made by trial and error with theoretical understanding coming later.

Four aspects of clock pendulums are discussed in this chapter. These are; pendulum quality, buoyancy, the effect of suspensions, and the effect of escapements. In the analysis of clock pendulum it is usually assumed that results from the analysis of an idealised compound pendulum (Sect. 3.2) and of simple harmonic motion (Sect. 2.2) can be applied, at least approximately, to real clock pendulums. Some further results from the analysis of simple harmonic motion are given in this chapter.

4.2 Pendulum Quality Q

In an idealised compound pendulum (Sect. 3.2) it is assumed that the only forces acting on its component parts are those due to gravity so, once launched, it will continue to swing indefinitely. In physics terminology an idealised compound pendulum is a non dissipative system. In a real compound pendulum inevitable frictional forces, and aerodynamic forces, retard the pendulum and mean that its oscillations will decay, and it will eventually stop. A real compound pendulum is a *dissipative system*. The retarding forces are relatively less important if a heavy bob is used. A heavy bob provides what clockmakers used to call '*dominium over the clock*' (Britten 1892, 1978). The bob in an eighteenth Century longcase clock with a seconds pendulum (Fig. 1.2b) weighs 750 g. The bob shown in Fig. 4.2 is from a high quality clock with a seconds pendulum. It is 21 cm long and weighs $5^1/_2$ kg. Clocks with short pendulums usually have light bobs. The bob in an early twentieth century wall clock (Fig. 4.1b) weighs 70 g. How rapidly the oscillations decay is a measure of *pendulum quality*. A widely used measure of pendulum quality is based on analysis of damped simple harmonic motion, and is given the symbol Q.

4.2.1 Damped Simple Harmonic Motion

Simple harmonic motion arises in systems with one degree of freedom where the restoring force F, towards a rest position is proportional to the distance from the rest

Fig. 4.2 Bob from a high quality clock with a seconds pendulum

position. A simple example (Sect. 2.2) is a point mass, m, which can move along a straight line with a restoring force towards a fixed point, O, that is proportional to the distance from the fixed point, x. This is a non dissipative system and, once launched, simple harmonic motion continues indefinitely. In any real physical system motion is always opposed by forces due to friction etc., and the amplitude diminishes with time.

In *damped simple harmonic motion* a resistance force, proportional to velocity, is opposed to the motion of the point mass, m, along the straight line and Eq. 2.1 becomes (Bateman 1977a)

$$m\frac{\mathrm{d}^2y}{\mathrm{d}t^2} = -F - R\frac{\mathrm{d}x}{\mathrm{d}t} \tag{4.1}$$

where t is time, F is the restoring force per unit distance, and R is the resisting force per unit velocity. Substituting Eq. 2.5 into Eq. 4.1 and taking

$$k = R/m \tag{4.2}$$

where k is the *damping constant*, leads to

$$\frac{\mathrm{d}^2x}{\mathrm{d}t^2} = k\frac{\mathrm{d}x}{\mathrm{d}t} + \omega^2 x = 0 \tag{4.3}$$

where ω is angular velocity. Making the further substitution

$$x = y\exp\left(\frac{-kt}{2}\right) \tag{4.4}$$

$$\frac{\mathrm{d}^2y}{\mathrm{d}t^2} + y\left(\omega^2 - \frac{k^2}{4}\right) = 0 \tag{4.5}$$

which is valid for $\omega > k/2$. Comparison with Eq. 2.11 leads to identification of the *damped angular velocity*, ω_d, given by

$$\omega_\mathrm{d} = \sqrt{\omega^2 - \frac{k^2}{4}} \tag{4.6}$$

Hence, the damped simple harmonic motion of the point mass, m, is given by (cf Eq. 2.3)

$$x = A\exp\left(\frac{-kt}{2}\right)\cos\left(\omega_\mathrm{d}t\right) \tag{4.7}$$

where A is an arbitrary constant, and is shown schematically in Fig. 4.3. The term exp(-$kt/2$) shows that the amplitude decays exponentially to zero. The time taken for the amplitude to decay from A to A/e is called the *time constant*, τ.

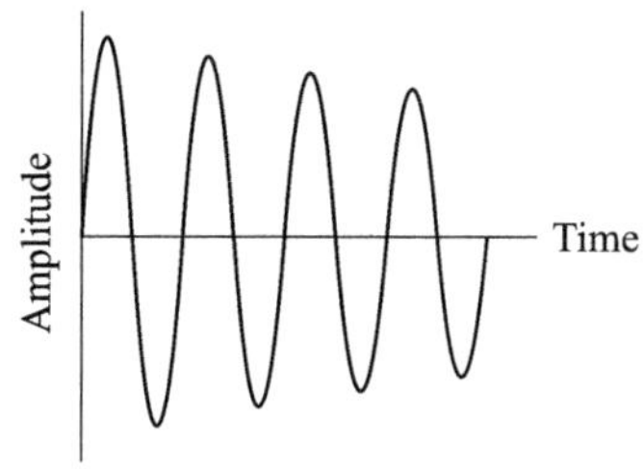

Fig. 4.3 Damped simple harmonic motion

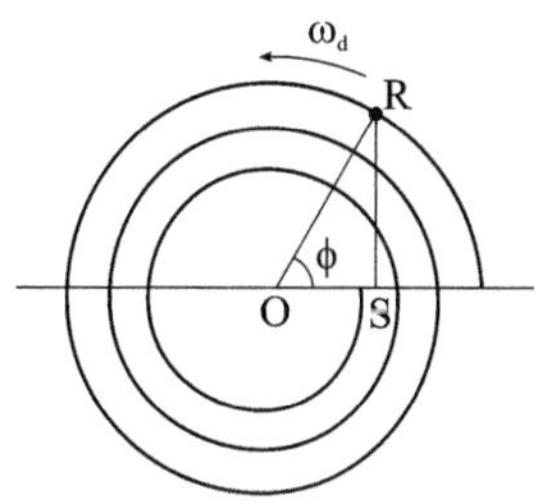

Fig. 4.4 Graphical representation of damped simple harmonic motion

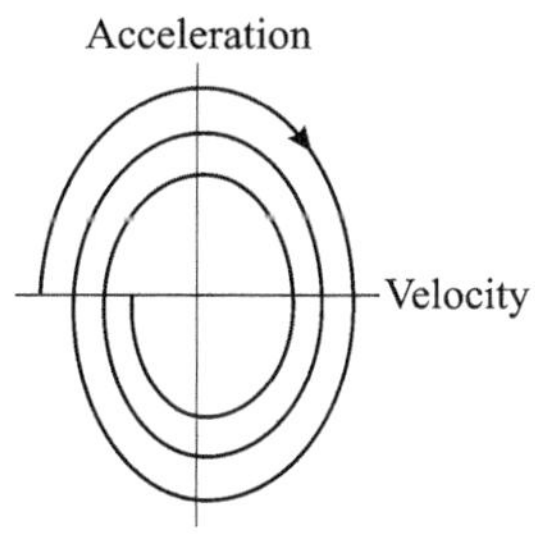

Fig. 4.5 Phase diagram for damped simple harmonic motion. The *arrow* shows the sense of increasing time

The natural logarithm of the ratio of successive peak amplitudes is the *logarithmic decrement*, δ, and is given by

$$\delta = T/\tau \tag{4.8}$$

It follows that

$$k = \delta/\pi \tag{4.9}$$

A graphical representation of damped simple harmonic motion is shown in Fig. 4.4 (cf Fig. 2.2). In the figure a point moves around an equiangular spiral (Coxeter 1961) with constant angular velocity, ω_d. When it is at point R the projection, point S, onto a horizontal line, gives the position of the point mass moving with damped simple harmonic motion along the horizontal line. The velocity of the point mass at point S is ω_dRS, and its phase is the angle ϕ. The phase diagram for damped simple harmonic motion is a spiral (Fig. 4.5).

4.2.2 Definition of Q

A convenient definition of Q is based on damped simple harmonic motion (previous section), and it is given by

$$Q = \pi/\delta \tag{4.10}$$

where δ is the logarithmic decrement. Other, equivalent, definitions of Q are possible and, for resonant systems in general, it is called the *quality factor*. The first use of the term appears to have been in 1938 (Bateman 1977b). When applied to pendulums Q is called *pendulum quality*.

The theory of damped simple harmonic motion is based on the assumption that a resistance force, proportional to the velocity of a point mass, m, is opposed to the motion of the point mass, m. An equivalent assumption for a *damped pendulum* is that a resistance torque, opposed to the motion of the pendulum, is proportional to the angular velocity of the pendulum. However, this is only roughly true for a real pendulum (Bacon and Nguyen 2005; Aggarwal et al. 2005) a point that is often overlooked. Nevertheless, Q does give an indication of the effect of changes made to a particular pendulum, but comparisons between different pendulums should be treated with caution.

In a real pendulum the effect of losses due to friction etc. can be represented by a *phase diagram* in which angular acceleration is plotted against angular velocity. This is a spiral (Fig. 4.6) similar to the phase diagram for damped harmonic motion (Fig. 4.5). In reality the amplitude of a pendulum decreases much more slowly than shown in Fig. 4.6. A practical definition of Q, equivalent to Eq. 4.10 is (Bateman 1977b; Matthys 2004)

$$Q = 4.532n \tag{4.11}$$

where n is the number of to and fro swings for the amplitude to decay to half its initial value. For a clock pendulum it is appropriate to take the initial amplitude as that during normal operation of the clock. As $Q \to \infty$ a pendulum becomes a non dissipative system and swings indefinitely once launched.

It is perhaps self evident that a clock pendulum whose amplitude decreases slowly is of better quality than one whose amplitude rapidly in that less energy is required to keep it swinging with constant amplitude. The use of Q as a measure

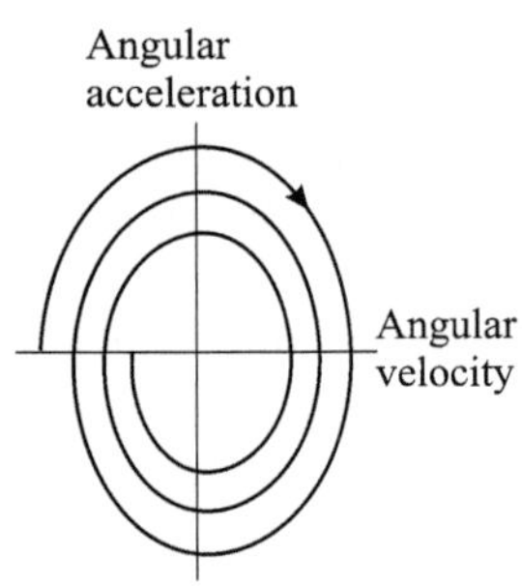

Fig. 4.6 Phase diagram for a real pendulum. The *arrow* shows the sense of increasing time

of pendulum quality is relatively recent Bateman (Bateman 1977b). It has the advantages that it is non dimensional and increases with increasing pendulum quality. Its use is now widespread, and the term Q is often used in the horological literature without explanation. There is a strong correlation between the accuracy of a clock and the Q of its pendulum Bateman (1977c), but Matthys (2004) points out that there is no convincing physical explanation for this correlation. He states that Q for clock pendulums is in the range 2,200–500,000. This apparently applies to high quality clocks since lower Q could be expected for poor quality clocks. Higher values of Q are only attainable at low pressure since this reduces losses due to air resistance.

In pendulum systems with more than one primary mode of oscillation oscillations tend to degenerate into the primary mode of oscillations with the highest Q (Haine 2010). This explains why motions of the *double rod pendulum* (Sect. 3.3) tend to degenerate into the mode of oscillation shown in Fig. 3.3a.

4.3 Buoyancy

Archimedes principle, discovered by Archimedes in 212 BC, states that when an object is immersed in a fluid there is an upward force on the object, known as *buoyancy*, that is equal to the weight of fluid displaced. Buoyancy due to immersion in air reduces the apparent weight of a pendulum but not its mass, and for brass the effect is to increase the pendulum period by about 0.007% (Rawlings 1993). Changes in barometric pressure lead to *barometric error*. For a clock with a brass pendulum bob this is about 0.006 s per day per millibar. Barometric error is usually negligible, but correction is needed in some precision clocks. Methods of correction were suggested by the Rev Dr Robinson in 1831 (Rawlings 1993). One approach is to use an aneroid capsule to move a weight such that the effective length of the pendulum is reduced as barometric pressure increases.

4.4 Suspensions and Modes of Oscillation

For accurate timekeeping, the essential requirements of a pendulum *suspension* are that it should allow the pendulum to swing as freely as possible in the desired mode of oscillation, and that any unwanted modes of oscillation should be suppressed as much as possible (Matthys 2004). In other words the pendulum quality, Q, for unwanted modes of oscillation should be much lower than for the desired mode of oscillation. Hence, it is the difference between Q for the desired mode of oscillation and unwanted modes of oscillation, rather than the absolute value of Q for the desired mode of oscillation, that is important. For clock pendulums the usual desired mode of oscillation is that the centre of oscillation of the pendulum moves in an arc in a vertical plane and, for small amplitudes, approximates to simple harmonic motion (Sect. 2.2). In Matthys' notation this is *left-right motion* as viewed from in front of a clock. The method of suspension of a real pendulum can only approximate to the frictionless pivot of an ideal pendulum.

There are three main types of pendulum suspension. The most widely used is the *spring suspension* in which the pendulum is suspended by a flat spring (Figs. 1.2a and 4.1b). A *pivot suspension* (Fig. 4.7c) is often used in cheap clocks. In a *knife edge suspension* a pair of hard knife edges are attached to the pendulum rod and these rest on hard horizontal beds (Fig. 4.8). For good timekeeping a pendulum must swing from a sufficiently rigid attachment (Britten 1978). This is especially the case when a heavy bob is used (Sect. 4.2).

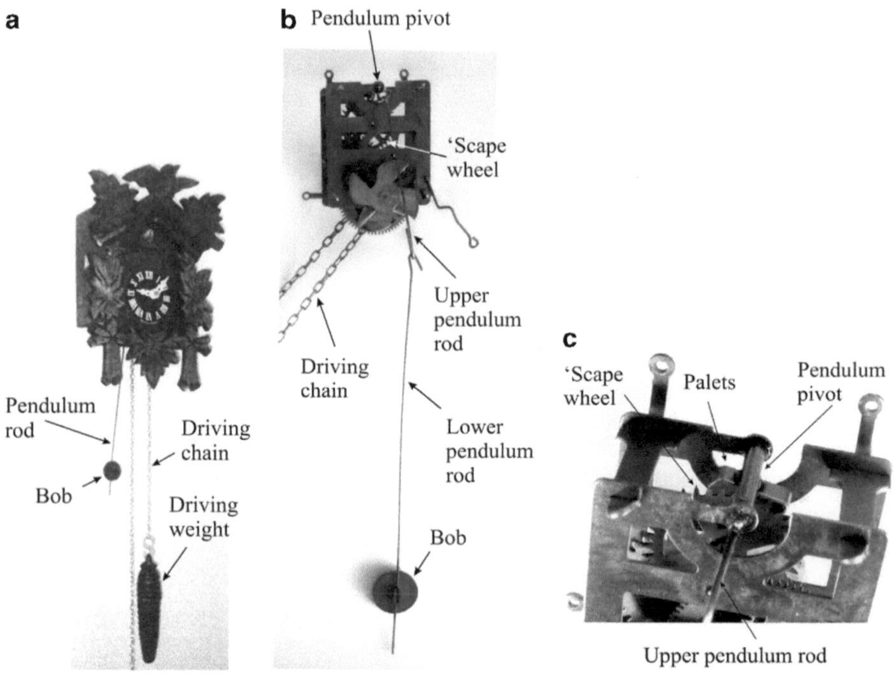

Fig. 4.7 Cuckoo clock with a pivot suspension. (**a**) General view. (**b**) Rear view of movement. (**c**) Pendulum pivot and escapement

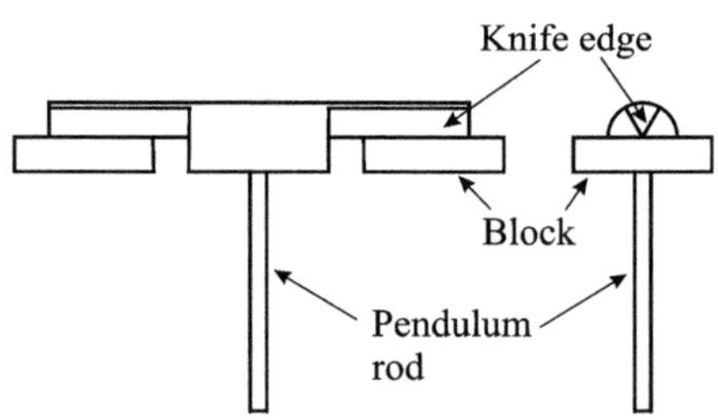

Fig. 4.8 Knife edge suspension

4.4.1 Spring Suspensions

The *spring suspension* (Figs. 1.2a, 4.1b and 4.9), in which the pendulum is suspended by a flat *suspension spring*, has been used for centuries. Some horological authors, for example Britten (1892) and Matthys (2004), do not mention any other type. The use of a flat spring in a allows the pendulum to swing freely in left-right motion in a vertical plane, which is the desired mode of oscillation. Detail design of the attachments of the suspension spring varies, but there is usually provision for *front-to-back motion* so that the pendulum can swing in a vertical plane in the desired mode of oscillation. Front-to-back motion, viewed from in front of a clock, is in a vertical plane perpendicular to that for left-right motion. A spring suspension pendulum is an example of a system that appears to be simple but is actually complex and difficult to analyse in detail.

A spring suspension pendulum has several degrees of freedom, so it is a chaotic system, and chaotic behaviour can occur. This can easily be demonstrated by launching a long case clock pendulum, such as that shown in Fig. 1.2b, in random directions. Suppression of unwanted modes of oscillation, and hence chaotic behaviour, requires careful attention to detail design (Matthys 2004). In other words, when a clock pendulum is oscillating in an unwanted mode of oscillation, the pendulum quality Q, should be low. Clocks with spring suspension pendulums usually have a crutch (Figs. 1.2a and 4.1b) to connect the pendulum to the escapement (Saunier 1861/1975; Britten 1978; Rawlings 1993). The crutch is attached to an arbor pivoted in the movement. This arrangement has been used for centuries: a pendulum clock movement including a crutch is illustrated in Huygens (1658).

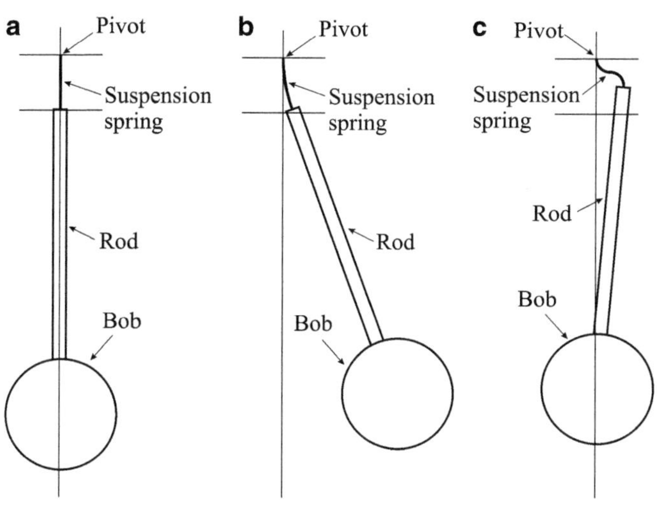

Fig. 4.9 Spring suspension pendulum. (**a**) Rest position. (**b**) Deflected, first primary mode of oscillation. (**c**) Deflected, secondary mode of oscillation

There are three primary modes of oscillation of a spring suspension pendulum (Matthys 2004). The first is the desired left-right motion using the suspension spring as an equivalent to a pivot. The second is unwanted front-to-back motion about a pivot, usually at the top of the suspension spring (Fig. 4.9). The third is unwanted rotation of the pendulum bob about the long axis of the pendulum. There is also an unwanted secondary mode of oscillation, in which there is left-right motion of the top of the pendulum rod, with up and down motion of the bob, but with little or no left-right motion of the bob (Fig. 4.9c).

In some small clocks the pendulum rod is in two parts connected by a hook (Fig. 4.1b) so that the lower pendulum rod and bob can easily be removed to facilitate moving the clock. The hook is arranged so that relative left-right motion between the two parts of the pendulum rod is not possible. In other words, the pendulum cannot oscillate as a real version of a double rod pendulum in left-right motion (Sect. 3.3). Theoretically it could oscillate as a real version of a double rod pendulum in front-to back motion, but in practice friction at the pivots means that the pendulum quality is too low for this to occur.

The first primary mode of oscillation is the desired mode of oscillation of a spring suspension pendulum. In this mode of oscillation the suspension spring curves into an arc, shown exaggerated in Fig. 4.9b, and there is a clearly defined axis of rotation of the pendulum below the top of the suspension spring. In this mode of oscillation the pendulum is analogous to the idealised compound pendulum shown in Fig. 3.1. For a suspension spring 15.24 mm long the axis of rotation was found to be 5.61 mm below the top of the spring (Rawlings 1993), that is about one third of the way down. Some measurements showed that the axis of rotation moves slightly downwards as amplitude increases (Matthys 2004). In other words the pendulum, effectively, becomes shorter, which would provide some compensation for circular error (Sect. 2.3.4). In the second subsidiary mode of oscillation the suspension spring is curved into an S-shape, shown exaggerated in Fig. 4.9c. In this mode the pendulum bob moves up and down, and the motion is analogous to the mode of oscillation of a double rod pendulum shown in Fig. 3.3c.

The combination of the first primary mode of oscillation and the second subsidiary mode of oscillation (Fig. 4.9b, c) has been analysed in detail by Emmerson (1999, 2010). He points out that, for small amplitudes, the displacement of the pendulum where it passes through the fork at the lower end of the crutch is the algebraic sum of the displacements due to each of the modes. The result is that the motion of the crutch fork is irregular and does not approximate to simple harmonic motion, (Fig. 2.5). As a qualitative indication of what might happen in an actual clock, Fig. 4.10 was constructed assuming that the crutch fork displacement due to the first primary mode of oscillation is simple harmonic motion with amplitude 1 and pendulum period 1 s, and that due to the second subsidiary mode of oscillation is simple harmonic motion with amplitude $1/3$ and pendulum period 0.615 s. The resulting combined motion of the crutch fork is chaotic. The combined motion is cyclic, but the amplitude varies from cycle to cycle, and the crutch fork displacement

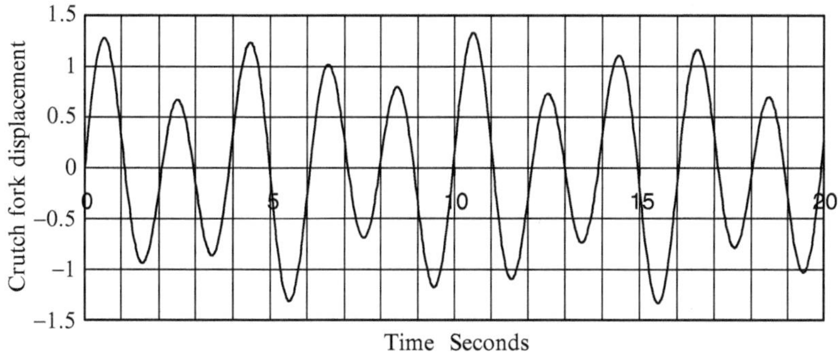

Fig. 4.10 Crutch fork displacement as the algebraic sum of displacements due to the first primary mode of oscillation and the second subsidiary mode of oscillation

does not pass through zero at intervals of exactly 1 s. The appearance of the figure is similar to one given by Emmerson (2010), which is based on calculations for a typical spring suspended pendulum with a pendulum period of 1 s.

4.4.2 Pivot Suspensions

A *pivot suspension* is often used in cheap pendulum clocks, but is not usually mentioned by horological authors, although Rawlings (1993) does give an example. In a pivot suspension, the pendulum rod is attached to an arbor pivoted in the movement. Figure 4.7 shows a cuckoo clock with a pivot suspension. In this particular clock the pendulum rod is in two parts with hooks formed at the lower end of the upper pendulum rod, and the upper end of the lower pendulum rod.

The major disadvantage of a pivot suspension is that, because of friction in the pivot it is only possible to use a light bob (Rawlings 1993) so timekeeping tends to be indifferent. The bob shown in Fig. 4.7b is made of plastic and only weighs a few grams. A pivot suspension permits left-right motion but not front-to back motion. The pendulum rod is therefore usually in two parts, connected by a hook, arranged so that the lower pendulum rod can move front-to-back and hence swing in a vertical plane. Ideally, the hook should be arranged so that relative left-right motion between the two parts of the pendulum rod is not possible. However, the hooks shown in Fig. 4.7b do allow left-right motion. The pendulum has three degrees of freedom, and can display chaotic behaviour.

The desired mode of oscillation for the cuckoo cock shown in Fig. 4.7 is left-right motion with no relative motion between the upper and lower pendulum rods, but this does not happen. Instead there are two primary modes of oscillation corresponding to modes of oscillation of a double rod pendulum (Sect. 3.3). In the first primary mode of oscillation the pendulum angles, θ_1 and θ_2, move in phase, corresponding to the mode of oscillation for a simple rod pendulum shown in Fig. 3.3a. In this mode of oscillation the clock runs at about the correct rate. There is relative motion between the upper and lower pendulum rods. In the second primary mode of oscillation, for which the frequency is higher, the pendulum angles, θ_1 and θ_2, move out of phase and the bob moves approximately up and down, corresponding to the mode of oscillation shown in Fig. 3.3c, and the clock runs far too fast. The second primary mode of oscillation tends to degenerate into the first primary mode of oscillation. In the second primary mode of oscillation there is a large amount of relative motion between the upper and lower pendulum rods The large amount of relative motion between the upper and lower pendulum rods means that the pendulum quality, Q, is lower than for the first primary mode of oscillation. This explains why the second primary mode of oscillation tends to degenerate into the first primary mode of oscillation (Sect. 4.4.2).

4.4.3 Knife Edge Suspensions

In a *knife edge suspension* a pair of hard knife edges are attached to the pendulum rod and these rest on hard horizontal beds, as shown schematically in Fig. 4.8 (Rawlings 1993). A knife edge usually has an included angle of about 60°. Similar knife edge suspensions are used in chemical balances.

Knife edge suspensions are sometimes used in precision clocks with one second pendulums (Roberts 2003). Friction is relied on to keep the suspension properly aligned. Detail design varies, but there is usually a pivot to allow front-to-back motion so that the pendulum can swing in left-right motion in a vertical plane, which is the desired mode of oscillation. The loading on a knife edge used in a suspension is usually in the range 2.3–4.4 kg per cm length of knife edge (Rawlings 1993; Bateman 1993). The knife edge has to have a small radius in order to avoid deformation or chipping (Johnson 1987; Rawlings 1993). In practice this does not cause difficulty.

In clocks with short pendulums, that are likely to be moved, the knife edges usually rest in shallow V-notches with an included angle of about 120° (Penman 1998). Appropriate guards keep a knife edge in position if the clock is moved with the pendulum in position. The radius at the bottom of the V-notch must be greater than that of the knife edge.

4.5 Effects of Escapements

One of the functions of an *escapement* (Sect. 4.1) is to transfer energy to a pendulum to replace energy lost due to friction etc. It does this by impulses whose timing is determined by the pendulum. Hence the pendulum has dominium over the clock (Sect. 4.2), and the escapement should interfere as little as possible with the oscillations of a pendulum. However, as energy is being transferred it is

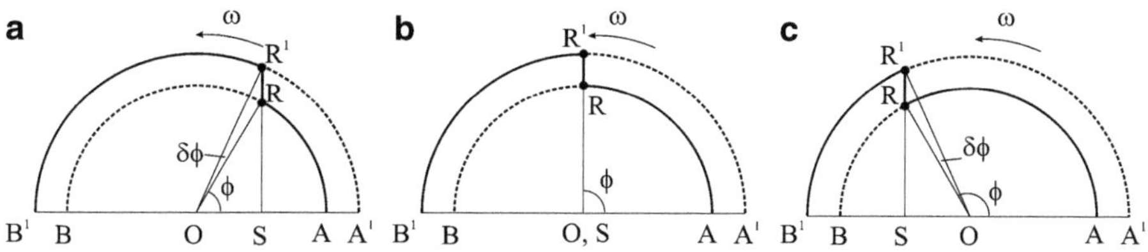

Fig. 4.11 Effect of an impulse in direction of motion on simple harmonic motion. (**a**) Before rest position. (**b**) At rest position. (**c**) After rest position

impossible to design an escapement which does not interfere with the oscillations of a pendulum. The effect of an escapement on the pendulum period is known as the *escapement error*. This is a misnomer, but it is the accepted term. Escapement errors can be either positive or negative and can either increase or decrease with pendulum amplitude (Tekippe 2010). A mechanical escapement (Figs. 1.2c, 1.3 and 4.7c) periodically applies a force to the pendulum over a short period of time, and these forces are called *impulses*. The effect of an instantaneous impulse on simple harmonic motion was first investigated theoretically by Airy in 1827 (Bateman 1977d; Rawlings 1993; Denny 2002). Airy's analysis is outlined by Denny (2002), and a modern derivation is given by Rawlings (1993).

Airy's results are illustrated by the graphical representations shown in Fig. 4.11. These show schematically what happens when an impulse is applied to the point mass, at point P, in its direction of motion. The notation used is based on that for the graphical representation of simple harmonic motion shown in Fig. 2.6 (Sect. 2.2). Before the impulse is applied the point R moves along the arc AR, and the vector OR moves with constant angular velocity, ω. The radius of the semi-circle, OA, is the initial amplitude of the oscillation, and ωRS is the velocity of the point mass at point S. The angle AOR is the *phase* of the point mass at S. When the impulse is applied the point R jumps vertically and becomes point R′. After the impulse is applied the vector OR′ moves with constant angular velocity, ω. The radius of the semi-circle, OA′, is the final amplitude of the oscillation, A, and ωR′S is the velocity of the point mass at point S.

There are three distinct cases. In the first, the impulse is applied while the point P is moving towards the rest position (Fig. 4.11a). When the impulse is applied the phase angle jumps from ϕ to $\phi + \delta\phi$. Apart from this jump the angular velocity is constant. The effect is to reduce the period of the half oscillation shown in the figure. In the second, the impulse is applied while the point P is moving through the rest position (Fig. 4.11b). The phase angle is unchanged so the period of the half oscillation shown in the figure is also unchanged. In the third, the impulse is applied while the point P is moving away from the rest position (Fig. 4.11c). When the impulse is applied the phase angle jumps from ϕ to $\phi - \delta\phi$ so the period of the half oscillation shown in the figure is increased.

In the horological literature it is usually assumed that, for small amplitudes, Airy's analysis applies to the effect of impulses from an escapement on the period of a pendulum oscillating in left-right motion, that in the first primary mode of oscillation (Sect. 4.4.1). However, in a practical escapement, acting on a real pendulum, the impulse is applied over a finite period of time so Airy's analysis, which is for an instantaneous impulse, is only an approximation. In general the impulses change the pendulum period, and this change is the *escapement error*. By analogy with Airy's analysis there are three distinct cases. In the first, an impulse is applied while the pendulum is moving towards the rest position and the effect is to reduce the pendulum period. In the second, an impulse is applied while the pendulum is at the rest position and the pendulum period is unchanged. In the third, an impulse is applied while the pendulum is moving away from the rest position and the pendulum period is increased. Airy's analysis implies that in order to minimise escapement error impulses should be applied while the pendulum is moving through the rest position or close to it. Despite Airy's result most improvements in escapement design have been the result of trial and error rather than theoretical input. Reviews of various types of escapement are given by Bateman (1977e), Britten (1978), Rawlings (1993) and Roberts (2003). Bateman (1977d) suggested that escapement error could be reduced by increasing the pendulum quality, Q, since this decreased the level of impulses needed to keep the pendulum in motion.

The effect of impulses on a pendulum can be represented graphically by a phase diagram in which angular acceleration is plotted against angular velocity. The phase diagram for a real pendulum driven by an escapement with impulses at the ends of the swing is shown schematically in Fig. 4.12a. The impulses are assumed to be short enough to be regarded as instantaneous. The effect of the impulses is exaggerated for clarity. One complete *circuit* of the phase diagram, in the direction shown by the arrow, represents one complete oscillation of the pendulum. Each impulse, represented by a horizontal line, causes an instantaneous increase in the angular velocity of the pendulum. Energy losses due to friction etc. mean that curved parts of the circuit are spirals (cf Fig. 4.6). If the impulses exactly replace the energy lost, then the circuit repeats exactly. In other words the pendulum amplitude is constant.

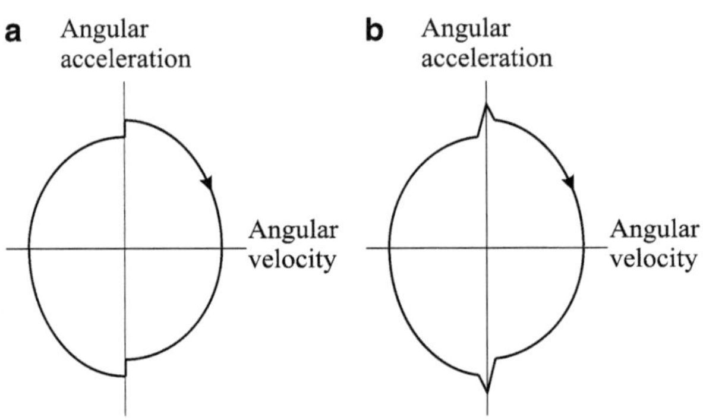

Fig. 4.12 Phase diagram for a real pendulum. The effect of the impulses is exaggerated for clarity. *Arrows* show the sense of time. (**a**) Driven by an instantaneous impulse at the end of a swing. (**b**) Driven by a recoil escapement with an impulse of finite length around the end of a swing

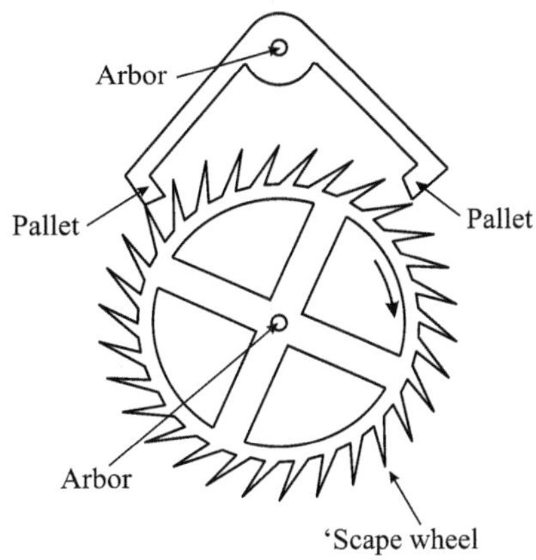

Fig. 4.13 Sketch of dead beat escapement

A pendulum clock driven by a constant force, such as that due to a *driving weight* (Figs. 1.2b and 4.7a), is a stable *dynamical system* (Denny 2002; Heldman 2009). In its stable state the phase diagram repeats exactly and is the corresponding *limit circuit*. If the pendulum amplitude is disturbed by an extraneous impulse then the phase diagram spirals back to the limit circuit. A numerical example is given by Denny (2002). When a clock is launched by pushing the pendulum it automatically settles into the stable state.

The *dead beat escapement* (Fig. 4.13) was first used successfully by George Graham in 1815, but he was probably preceded by Thomas Tompion (Britten 1978). Edwardes (Edwardes 1977) suggests that Graham's escapement was probably an improvement of earlier versions. It is sometimes called the *Graham escapement*, and is used in precision pendulum clocks (Roberts 2003, 2004). The teeth are of a different shape to those of an anchor escapement (Fig. 1.3) and are such that the 'scape wheel rotates intermittently, always in the same direction. It simply stops at each tick of the clock, hence the term 'dead beat'. Since 'scape wheels do not move continuously they are *crossed out* to make them lighter and hence reduce inertia effects. Impulses applied by a dead beat escapement are fairly close approximations to instantaneous impulses.

In a clock fitted with a *recoil escapement* the 'scape wheel reverses during an impulse. If the clock is fitted with a seconds hand then the seconds hand visibly recoils for a short distance after each tick. The *anchor escapement* (Figs. 1.2c and 1.3) is a widely used and reliable recoil escapement that was invented about 1670 (Britten 1978). The invention of the anchor escapement is sometimes attributed to Robert Hooke (Britten 1892; Bennet et al. 2003) but this has not been verified Roberts (2003). It has also been attributed to William Clement (Edwardes 1977).

The phase diagram for a real pendulum driven by a recoil escapement is shown schematically in Fig. 4.12b. The impulse is of finite length and extends around the end of a swing. The force due to the impulse is always in the same direction, but the direction of motion of the pendulum reverses at the end of a swing. Hence, before the end of the swing the force due to the impulse is in the same direction as the motion of the pendulum and the angular acceleration of the pendulum increases, but after the end of the swing the force due to the impulse is in the opposite direction to the motion of the pendulum and the angular acceleration decreases. The net effect can be represented as an instantaneous impulse at the end of a swing, as shown in Fig. 4.12a.

The numerous variables make it impossible to derive precise theoretical values of the effect of escapement error on the pendulum period, and a trial and error approach still has to be used. Modern precision electronic timing equipment makes it possible to measure pendulum periods very accurately. This makes a trial and error approach easier since the effects of changes made to an escapement can found quickly and easily (Tekippe 2010).

4.6 Reduction of Effects of Circular Error

What is known as *circular error* (Sect. 2.3.4) means that the period of a pendulum increases as its amplitude increases. The amplitude of a clock pendulum cannot be kept precisely constant so minimising changes in pendulum period due to inevitable changes in amplitude is an important aspect of clock design. Unwanted changes in amplitude are more significant for a pendulum swinging with a large amplitude, so precision clocks usually have small amplitude pendulums (Roberts 2003, 2004). For large pendulum amplitude there are two ways to reduce changes in pendulum period due to circular error. One is to reduce the effective length of a pendulum as the pendulum angle, θ, (Fig. 3.1) increases. The other is to increase the restoring force as the pendulum angle increases.

Huygens (Huygens 1658) showed that if the path of the point mass. m, of a *simple string pendulum* (Fig. 2.1b) is a *cycloid*, then the pendulum is isochronous, that is there is no circular error. To achieve this he proposed the use of cycloidal cheeks to make a pendulum isochronous. This is shown in Fig. 4.14 for a simple string pendulum (Sect. 2.4), swinging in a vertical plane. As desired, the effective length of a pendulum is reduced as the pendulum angle, θ, increases.

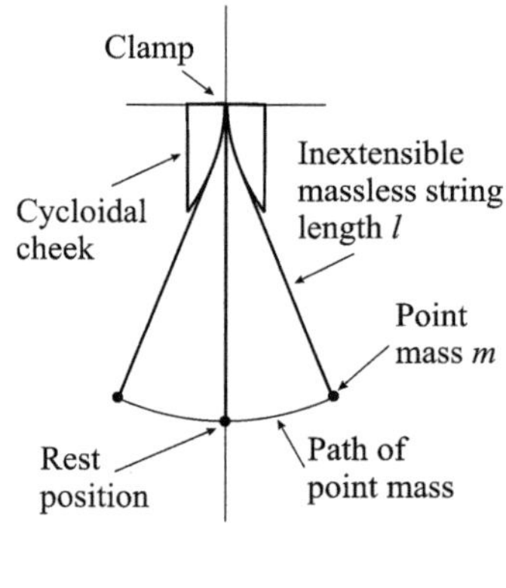

Fig. 4.14 Simple string pendulum with cycloidal cheeks

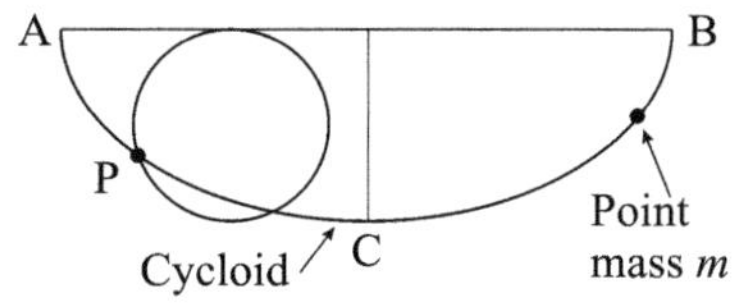

Fig. 4.15 Generation of a cycloid by a rolling circle

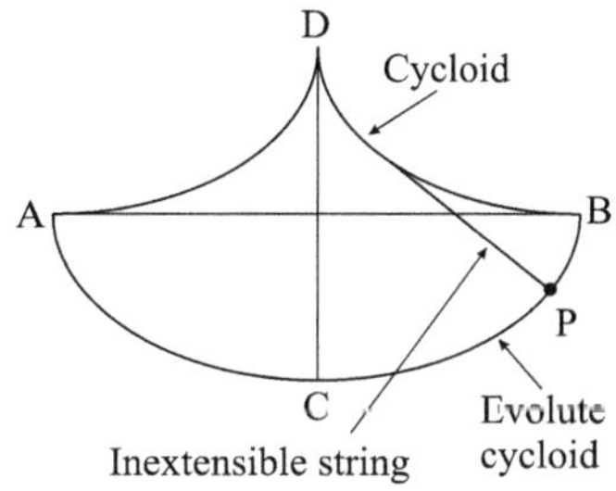

Fig. 4.16 Generation of a cycloid as an evolute of an identical cycloid

A cycloid is the path traced by a point P on the circumference of a circle when it is rolled along a straight line AB, as shown in Fig. 4.15 (Rawlings 1993). The arc ACB in the figure corresponds to one complete revolution of the circle. Rolling the circle past A or B results in a cusp, such as that at D in Fig. 4.16. When a point P moves along a curve the centre of curvature of the curve moves along another curve called the *evolute* which can be expressed parametrically in terms of its arc length (Coxeter 1961). The evolute of a cycloid is an identical cycloid, which can be expressed graphically as shown in Fig. 4.16. An inextensible string, length equal to the arc length BD, is clamped at one at D, and held tight against the cycloid at the other end P. P then traces the desired identical evolute cycloid.

The same idea can be used for a real pendulum fitted with a spring suspension (Sect. 4.4.1). In practice very few clocks are fitted with Huygens' cycloidal cheeks since it has been found that their use does not result in a significant improvement in timekeeping (Rawlings 1993). This is because of practical difficulties in persuading a real pendulum to follow a cycloidal path (Emmerson 2005).

The use of *auxiliary springs* to increase the restoring force as the pendulum angle increases was suggested by Edouard Phillips in 1887 (Roberts 2004). Restoring springs are used in Bulle clocks (Miles 1995; Bavister 2008), which were produced in the first quarter of the twentieth Century (Rawlings 1993). Bulle clocks are battery powered and usually have a pivot suspension (Sect. 4.4.2). Impulses are given to the pendulum by a coil mounted on the pendulum which swings over a permanent magnet mounted on the clock case (Fig. 4.17). The coil acts as a bob and there is a rating nut below it. A contact near the top of the pendulum completes the circuit to provide the impulses (Robinson 1942).

Pendulums in Bulle clocks have a large amplitude, and an auxiliary spring is used to compensate for changes in pendulum period due circular error. In this context an auxiliary spring is called an *isochronous spring*. Figure 4.18 shows how the isochronous spring is arranged on a Bulle pendulum with a pivot suspension. The

Fig. 4.17 Bulle clock with electromagnetically driven pendulum

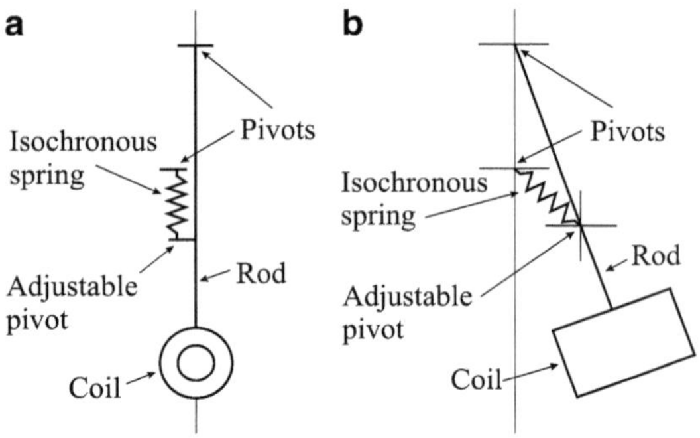

Fig. 4.18 Arrangement of the isochronous spring on a Bulle pendulum with pivot suspension. (**a**) View from side in rest position. (**b**) View from front away from rest position

upper end of the synchronous spring is attached to a fixed pivot, and the lower end to an adjustable pivot on the pendulum rod. In the rest position (Fig. 4.18a) there is some tension in the isochronous spring. As the pendulum swings away from the rest position (Fig. 4.18b) the length of the isochronous spring increases, and the increasing tension provides increasing additional restoring force as the amplitude increases. The synchronous spring is very light so its pivots introduce negligible additional friction. Edouard Phillips analysed the synchronous spring using elliptic integrals, and his analysis was published posthumously in 1898 (Bavister 2008). Numerical calculations by Bavister, using values taken from a Bulle clock, showed that with careful adjustment of the adjustable pivot, a synchronous spring could be very effective in reducing changes in pendulum period due to circular error. This was confirmed by measurements of the pendulum period on the same Bulle clock (Ridout and Thackery 2008).

Another approach is to compensate for changes in pendulum period due to circular error by ensuring that changes in pendulum period due to *escapement error* (previous section) are of opposite sign (Tekippe 2010). He states that in 1763 Ferdinand Berthoud knew that he could make a clock that kept better time by modifying the escapement (Tekippe 2010). Berthoud carried out an experiment with three different escapements driving the same pendulum. An escapement with a large recoil decreased the pendulum period when the driving weight, and hence the pendulum amplitude, was increased. A dead beat escapement increased the pendulum period when the driving weight was increased. An escapement with a little recoil caused no change in the pendulum period when the driving weight was increased. In other words it was isochronous for changes in amplitude. This concept of compensating circular error by changes in escapement error of opposite sign became part of traditional knowledge among French clockmakers. Saunier (1861/1975) was clearly familiar with the idea.

Experiments reported by Tekippe (2010) using modern precision electronic timing equipment and an escapement with a slight recoil showed clearly what happened when the driving weight was doubled. The pendulum amplitude increased until it reached the new stable value (limit cycle) corresponding to the increased weight. There was an initial step decrease in the pendulum period due the change in escapement error. The pendulum period then increased as the pendulum amplitude increased, initially rapidly, until it settled to a value corresponding to the increased driving weight. In this particular experiment the stable value of the pendulum period was unchanged by increasing the driving weight. In other words there is an isochronous combination of pendulum and escapement.

References

Aggarwal N, Verma H, Arun P (2005) Simple pendulum revisited. Eur J Phys 36(3):517–523

Anonymous (2010) Clock. http://en.wikipedia.org/wiki/Clock Accessed 10 Feb 2010

Bacon ME, Nguyen DD (2005) Real-world damping of a physical pendulum. Eur J Phys 26(4):651–655

Bateman D (1977a) Vibration theory and clocks. Part 1. Harmonic motion. Horological J 120(1): 3–8

Bateman D (1977b) Vibration theory and clocks. Part 2. Forced harmonic motion. Horological J 120(2):48–52

Bateman D (1977c) Vibration theory and clocks. Part 3. Q and the practical performance of clocks. Horological J 120(3):48–55

Bateman D (1977d) Vibration theory and clocks. Part 4. Q and classical escapement theory. Horological J 120(4):66–70

Bateman D (1977e) Vibration theory and clocks. Part 6. Errors in escapements. Horological J 120(6):65–73

Bateman D A (1993) Knife edges. Horological Sci Newsl (1993–1):5

Bavister R (2008) A study into the effectiveness of the isochronous spring. Electrical Horology Group Paper No. 76. Antiquarian Horological Society, Ticehurst

Bennet J, Cooper M, Hunter M, Jardine L (2003) Londons Leonardo. The life and work of Robert Hooke. Oxford University Press, Oxford

Britten FJ (1892) The watch & clockmakers' handbook, dictionary and guide, 8th edn. E & FN Spon, London

Britten FJ (1978) The watch & clock makers' handbook, dictionary and guide, 16th edn. Arco Publishing Company, New York (Revised by Good R)

Coxeter HSM (1961) Introduction to geometry. Wiley, New York

Denny M (2002) The pendulum clock: a venerable dynamical system. Eur J Phys 23(4):449–458

Edwardes EL (1977) The story of the pendulum clock. John Sherratt & Son Ltd., Altringham

Edwardes EL (1981) The suspended foliot and new light on early pendulum clocks. Antiquarian Horology, 12(6): 614–626, 634

Emmerson A (1999) Some mathematics of the cantilever pendulum. Horological Sci Newsl, (1999–3), 11–19 (Errata. Ibid., (1999–4), 26)

Emmerson A (2005) Things are seldom what they seem. Christiaan Huygens, the pendulum and the cycloid. Horological Sci Newsl 2005–1:2–32

Emmerson A (2010) Equations of motion for the spring suspended pendulum. Horological Sci Newsl 2010–1:2–14

Haine J (2010) Synchronous oscillations of asymmetric coupled pendulums. Horological Sci Newsl 2010–5:6–13

Heldman AW (2009) The spring-barrel/fusee/anchor-recoil-escapement pendulum clock as an integrated system with phase-locked negative feedback. Horological Sci Newsl 2009–3:2–25

Huygens C (1658/1977) The horologium. Latin facsimile and English translation. In: Edwardes EL The story of the pendulum clock. John Sherratt & Son, Altringham, pp 60–97

Johnson KL (1987) Contact mechanics. Cambridge University Press, Cambridge

King A (2009) The pendulum clock. Antiquarian Horology 31(6):845–846

Matthys R (2004) Accurate clock pendulums. Oxford University Press, Oxford

Miles HA (1995) The Bulle-clock of Favre-Bulle. Practical manual for the use of clockmakers and jewellers. Antiquarian Horological Soc, Ticehurst (Trans)

Penman L (1998) Practical clock escapements. Mayfield, Ashbourne

Penney D (2009) The earliest pendulum clocks: a re-evaluation. Antiquarian Horology 31(5): 614–620

Rawlings AL (1993) The science of clocks and watches, 3rd edn. British Horological Institute Ltd., Upton

Ridout M, Thackery S (2008) Addendum to EHG Paper No. 76. In: Bavister RA study into the effectiveness of the isochronous spring. Electrical Horology Group Paper No. 76. Antiquarian Horological Society, Ticehurst

Roberts D (2003) Precision pendulum clocks: 300 year quest for accurate timekeeping in England. Schiffer Publishing, Atglen

Roberts D (2004) Precision pendulum clocks: France, Germany, America, and recent advancements. Schiffer Publishing, Atglen

Robinson TR (1942) Modern clocks. Their repair and maintenance, 2nd edn. N A G Press Ltd., London

Saunier C (1861/1975) Treatise on modern horology. W & G Foyle Ltd., London. (Trans by Tripplin J, Rigg, E)

Tekippe B (2010) A simple regulator. Clock Watch Bull 53(385):131–138

Chapter 5
Driven Pendulums

5.1 Introduction

One of the functions of the escapement in a pendulum clock is to transfer energy to the pendulum in order to replace energy lost due to friction etc. It does this by impulses whose timing is determined by the pendulum (Sect. 4.5). By contrast, in a *driven pendulum*, energy is transferred to a pendulum by periodic forces whose period is not controlled by the pendulum. A driven pendulum is sometimes called a *forced pendulum*.

There are three distinct ways in which a *simple rod pendulum* (Sect. 2.3) can be driven. In *rotary driving* the pendulum is subjected to prescribed varying torques about the suspension point. Theoretical investigation of a damped simple rod pendulum under sinusoidally varying torque has shown that a wide range of chaotic behaviour sometimes occurs (D'Humieres et al. 1982; Rössler et al. 1990; Butikov 2008), especially for large amplitudes and low values of the pendulum quality, Q. Experimental results, using a specially designed real pendulum with an electromagnetic drive have confirmed the existence of both periodic and chaotic behaviour (Baker and Gollub 1996). A simple rod pendulum can be started from the rest position by using rotary driving. A simple string pendulum (Sect. 2.4) cannot be driven by rotary driving.

In *horizontal driving* the suspension point of a simple rod pendulum is subjected to prescribed varying horizontal displacements (Lamb 1923; Tritton 1986, 1992). In *vertical driving*, sometimes called *lifting*, the suspension point is subjected to prescribed varying vertical displacements (Pippard 1988; Bishop et al. 1996). In both cases there are two forces on the simple rod pendulum. One is that due to gravity on the point mass, m, (Fig. 2.1a) and is constant. The other is that needed to impose the prescribed force or displacement, and this force varies with time. A simple rod pendulum can be started from the rest position by using horizontal driving, but it cannot be started from the rest position by using vertical driving. A simple string pendulum can be driven by horizontal driving, and by vertical driving.

L.P. Pook, *Understanding Pendulums: A Brief Introduction*, History of Mechanism and Machine Science 12, DOI 10.1007/978-94-007-1415-1_5,

A *prescribed force* or a *prescribed displacement* can be either periodic, such as a sine wave (Fig. 2.5) or random. In the latter case *random process theory*, described in this chapter, is needed for understanding the effects of driving. *Driven damped simple harmonic motion* provides a useful approximation to some aspects of the behaviour of driven pendulums. Some aspects of horizontal driving are described in this chapter.

5.2 Random Process Theory

There is no generally accepted definition of precisely what is meant by a *random process*. Broadly, a random process can be defined as a process which has a random appearance, such as those shown in Figs. 5.1a and 5.2. Alternatively, a random process can be defined as a process that displays chaotic behaviour. In a *Gaussian random processes* instantaneous values follow the *Gaussian distribution* (or *Normal distribution*). Many naturally occurring random processes, such as water waves, are approximately Gaussian. *Random process theory* (Papoulis 1965; Bendat and Piersol 2000; Pook 2007) can be used to analyse Gaussian random processes

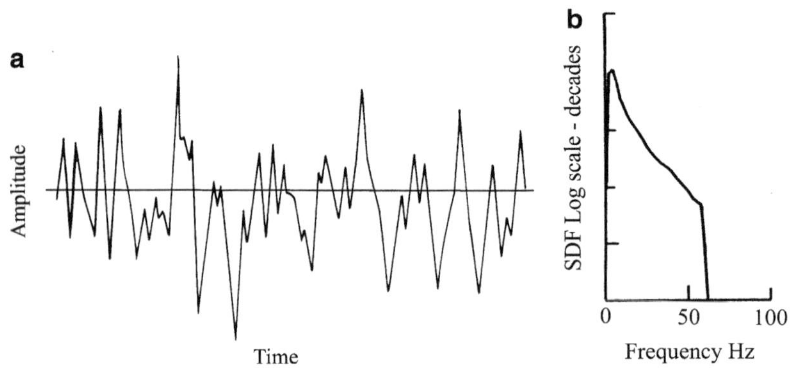

Fig. 5.1 Broad band random process, irregularity factor 0.410, spectral bandwidth 0.912. (**a**) Time history. (**b**) Spectral density function (Pook 1987) (Reproduced under the terms of the Click-Use Licence)

Fig. 5.2 Narrow band random process irregularity factor ≈0.99, spectral bandwidth ≈0.14 (Pook 1987) (Reproduced under the terms of the Click-Use Licence)

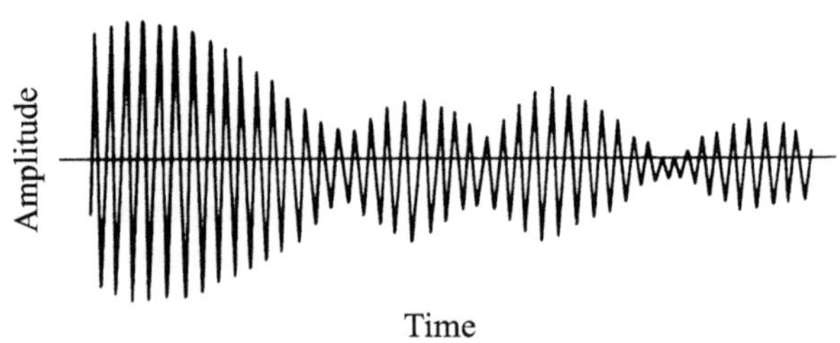

without the need to understand the processes. The random processes shown in Figs. 5.1a and 5.2 are both Gaussian random processes. They were computer generated so were precisely determined by the algorithms used, and can therefore be described as chaotic. In some ways random process theory is an alternative to chaos theory as described, for example by Baker and Gollub (1996), that can be used to highlight different aspects of behaviour. Figure 5.1a shows what is known as a *broad band random process*. In the figure amplitude is plotted against time. A feature of a broad band random process is that individual cycles cannot be distinguished. In a *narrow band random process* (Fig. 5.2) individual cycles can be distinguished and peaks have a slowly varying random amplitude.

In general, a random process may be described by the function $S(t)$, where S is amplitude and t is time. Assume that $S(t)$ is *statistically stationary* and *ergodic*. Stationary means that statistical parameters characterising the process are independent of time. Ergodic means, broadly, that different samples of the same process yield the same values for statistical parameters. Only stationary random processes can be ergodic, and in practice most are. Considering the time interval 0 to T the mean value of S, S_m is given by

$$S_m = \lim(T \to \infty) \frac{1}{T} \int_0^T S(t)\,dt \tag{5.1}$$

and the mean square value ϕ^2 by

$$\phi^2 = \lim(T \to \infty) \frac{1}{T} \int_0^T S^2(t)\,dt \tag{5.2}$$

Hence the *root mean square* (RMS) value, ϕ, is given by

$$\phi = \lim(T \to \infty) \sqrt{\frac{1}{T} \int_0^T S^2(t)\,dt} \tag{5.3}$$

The positive square root is understood in Eq. 5.3 and subsequent equations. The RMS can equally well be calculated for *periodic processes* such as a sine wave (Fig. 2.5). The use of RMS first became popular in electrical engineering because it can be used directly in calculations involving power. Carrying out calculations from the mean rather than from zero gives the *variance*, σ^2, where

$$\sigma^2 = \lim(T \to \infty) \frac{1}{T} \int_0^T \{S(t) - S_m\}^2\,dt \tag{5.4}$$

and the *standard deviation*, σ, is given by

$$\sigma = \lim (T \rightarrow \infty) \sqrt{\frac{1}{T} \int_0^T \{S(t) - S_m\}^2 \, dt} \tag{5.5}$$

The quantities given by Eqs. 5.1, 5.2 and 5.4 are related through the expression

$$\phi^2 = \sigma^2 + S_m^2 \tag{5.6}$$

For zero mean the RMS and standard deviation are numerically equal.

Instantaneous values of $S(t)$ may be characterized by *probability distribution functions*. These can be characterised by the *exceedance*, $P(S)$, which is the probability that a value exceeds S. The *cumulative probability*, $1 - P(S)$, is the proportion of values up to S. The *probability density*, $p(S)$, is the derivative of the cumulative probability. For convenience, S is often normalised by σ. The instantaneous values of many 'naturally occurring' random processes are statistically stationary, at least in the short term, and approximate to the Gaussian distribution which theoretically extends from $-\infty$ to $+\infty$. The probability density, $p(S/\sigma)$ of a Gaussian distribution (Fig. 5.3a) for a process with zero mean is given by

$$p\left(\frac{S}{\sigma}\right) = \frac{1}{\sqrt{2\pi}} \exp\left(\frac{-S^2}{2\sigma^2}\right) \tag{5.7}$$

and the exceedance (Fig. 5.3b) by

$$P\left(\frac{S}{\sigma}\right) = \frac{2}{\sqrt{2\pi}} \int_{S/\sigma}^{\infty} \exp\left(\frac{-S^2}{2\sigma^2}\right) d\left(\frac{S}{\sigma}\right) \tag{5.8}$$

This integral does not have an explicit solution. The values shown in Fig. 5.3b are for the positive half of a Gaussian distribution and are therefore twice those given by Eq. 5.8. $P(S/\sigma)$ is the area under the curve of $p(S/\sigma)$ between (S/σ) and infinity, as indicated by the shaded area in Fig. 5.3a.

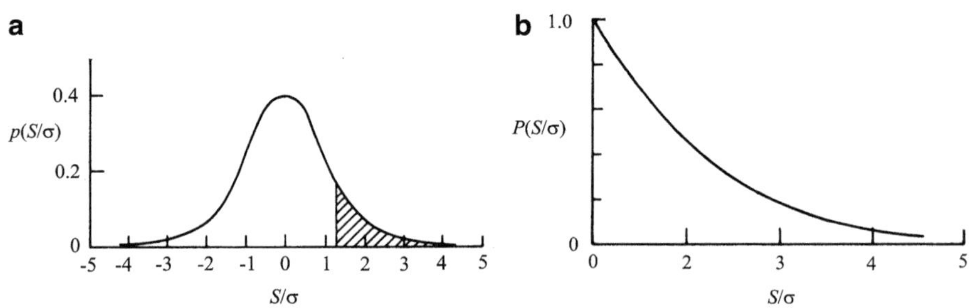

Fig. 5.3 Gaussian distribution. (**a**) Probability density. (**b**) Exceedance (Frost et al. 1974)

5.2.1 Bandwidth

The broad band random process shown in Fig. 5.1a, and the narrow band random process shown in Fig. 5.2 are both Gaussian random processes in that their instantaneous values follow the Gaussian distribution. The obvious difference between them can be expressed numerically by measures of what is called their *bandwidth*. Measures of bandwidth include the *irregularity factor*, the *spectral density function*, and the *spectral bandwidth*. The irregularity factor, I, is used in metal fatigue (Pook 2007). For a process with zero mean it is the ratio of upward going zero crossings to positive peaks, and lies in the range 0–1. The irregularity factor has the advantages that is easily understood, and is not restricted to Gaussian processes. The irregularity factor is 0.410 for the broad band random process (Fig. 5.1a) and about 0.99 for the narrow band random process (Fig. 5.2).

A Gaussian random process which is statistically stationary and ergodic can be described more precisely by its spectral density function (SDF) which describes the frequency content of the process. Mathematically, the SDF is obtained by first calculating the *autocorrelation function*, $R(\tau)$. This describes the relationship between the values of the random process $S(t)$ at times t and $t + \tau$, and is given by (Papoulis 1965; Bendat and Piersol 2000)

$$R(\tau) = \lim (T \to \infty) \frac{1}{T} \int_0^T S(t)\, S(t + \tau)\, \mathrm{d}t \tag{5.9}$$

where T is total time. The SDF, $G(f)$, where f is frequency, is the *Fourier transform* of $R(\tau)$, and is given by

$$G(f) = 2 \int_{-\infty}^{\infty} R(\tau) \exp(-i 2\pi f \tau)\, \mathrm{d}\tau = 4 \int_{-\infty}^{\infty} R(\tau) \cos(2\pi f \tau)\, \mathrm{d}\tau \tag{5.10}$$

Determining the SDF in this way is called transforming from the *time domain* to the *frequency domain*. The SDF is sometimes plotted on a logarithmic scale and sometimes on a linear scale. Figure 5.1b shows the SDF for the broad band random process shown in Fig. 5.1a. In practice it is usually calculated using an algorithm known as the *Fast Fourier Transform* (Bendat and Piersol 2000).

Some useful results depend only on the *spectral bandwidth*, ε, which is a measure of the RMS width of the SDF (Pook 1978; Bendat and Piersol 2000). It lies in the range 0–1 and is given by

$$\varepsilon = \sqrt{1 - \frac{m_2^2}{m_0 m_4}} \tag{5.11}$$

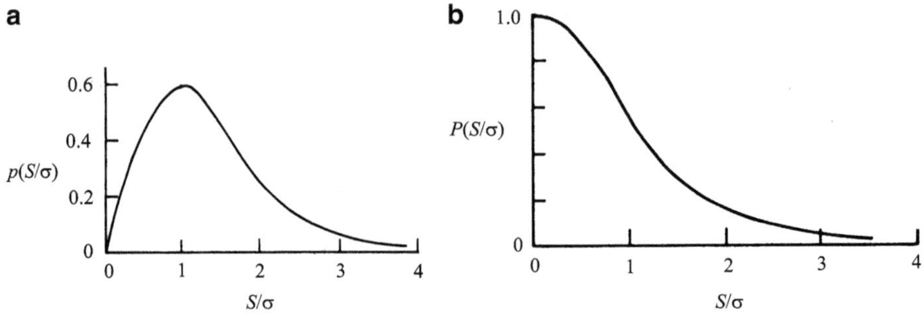

Fig. 5.4 Rayleigh distribution. (**a**) Probability density. (**b**) Exceedance (Frost et al. 1974)

where m_0, m_2 and m_4 are the zeroth, second and fourth moments of the SDF about the origin. Values of spectral bandwidth are shown in the captions to Figs. 5.1 and 5.2. It is related to the irregularity factor, I, by

$$\varepsilon^2 = 1 - I^2 \tag{5.12}$$

As $\varepsilon \to 0$ a Gaussians random process becomes the special case of a *narrow band random process* in which there are individual sinusoidal cycles (Fig. 5.2). These have a slowly varying random amplitude. There is no generally accepted definition of precisely what is meant by a narrow band random process, partly because of the physical difficulties of measuring bandwidth as $\varepsilon \to 0$ (Pook 2007).

The probability density for the occurrence of a positive peak of amplitude S (Fig. 5.4a) tends to the *Rayleigh distribution*, which is given by

$$p\left(\frac{S}{\sigma}\right) = \frac{S}{\sigma}\exp\left(\frac{-S^2}{2\sigma^2}\right) \tag{5.13}$$

where $p(S/)$ is the normalised probability density, and σ is the standard deviation of the process.

As the process is statistically symmetrical about zero, corresponding negative peaks also appear. The normalised exceedance, $P(S/\sigma)$, (Fig. 5.4b) is given by

$$P\left(\frac{S}{\sigma}\right) = \exp\left(\frac{-S^2}{\sigma^2}\right) \tag{5.14}$$

As a narrow band random process is Gaussian, the instantaneous values follow the Gaussian distribution (Eqs. 5.7 and 5.8).

Conventionally, in discussion of the Rayleigh and related distributions, only positive peaks are described and shown in diagrams such as Fig. 5.4, it being understood that the negative peaks, with due attention to sign, are also included. Negative peaks are sometimes called *troughs*. Theoretically the Rayleigh distribution extends to

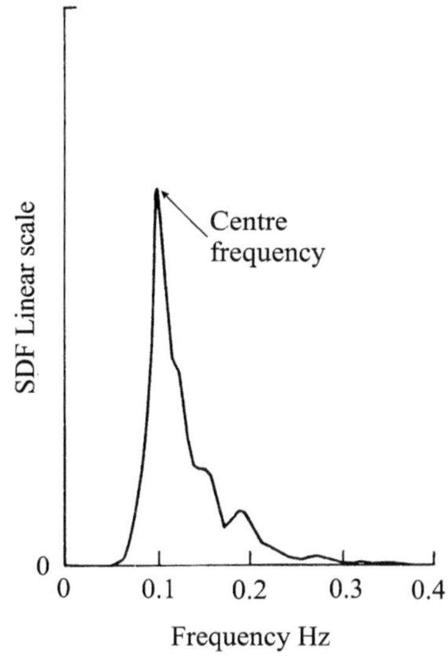

Fig. 5.5 Spectral density function for water surface elevation, significant wave height 4.75 m. (Pook 1989) (Reproduced under the terms of the Click-Use Licence)

infinity, but in practice peaks do not exceed a cut off value of S/σ, known as the *clipping ratio*. *Clipping* implies that higher peaks are reduced to the level given by the clipping ratio; *truncation* that they are omitted altogether. Physical limitations mean that the clipping ratio does not usually exceed four or five.

In a narrow band random process the spectral density function is sharply peaked at the *centre frequency*, as shown in Fig. 5.5. The figure is for water surface elevation of surface waves in the North Sea, significant wave height 4.75 m. The *significant wave height* is the average height, measured peak to trough, of the highest one third portion of the waves (Pook 2007) so these are large waves. In general, large surface waves are a narrow band random process.

5.3 Driven Damped Simple Harmonic Motion

Simple harmonic motion arises in systems with one degree of freedom where the restoring force F, towards a rest position is proportional to the distance from the rest position. A simple example (Sect. 2.2) is a point mass, m, which can move along a straight line with a restoring force towards a fixed point, O, that is proportional to the distance from the fixed point, x. In *damped simple harmonic motion* (Sect. 4.2.1) a resistance force, proportional to velocity, is opposed to the motion of the point mass, m, along the straight line. In *driven damped simple harmonic motion* there is, in addition, a prescribed force on the point mass, m, along the straight line. This prescribed force can be either periodic or random.

5.3.1 Periodic Driving

In *periodic driving* a *damped simple harmonic motion* is driven by a prescribed periodic force. If this periodic force, frequency f_p, is represented by $F_p \cos \omega_p t$ where ω_p is the angular velocity ($=2\pi f_p$) of the periodic force (Eq. 5.5), and t is time, then Eq. 4.3 becomes (Bateman 1977)

$$\frac{d^2x}{dt^2} = k\frac{dx}{dt} + \omega^2 x = \frac{F_p}{m}\cos \omega_p t \tag{5.15}$$

where m is the point mass. The steady state amplitude, A, of the oscillations, after any transient effects have decayed is given by

$$A = \frac{\frac{F_p}{m}}{\sqrt{\left(\omega^2 - \omega_p^2\right)^2 + \left(k\omega_p\right)^2}} \tag{5.16}$$

As the damping. $k \to 0$, $(\omega^2 - \omega_p^2) \to 0$ so from the equation the amplitude, $A \to \infty$. The phase, ε, between the driving force and the oscillation is given by

$$\tan \varepsilon = \frac{k\omega_p}{\omega^2 - \omega_p^2} \tag{5.17}$$

This analysis of *driven damped simple harmonic motion* leads to several equivalent definitions of the quality factor, Q (Bateman 1977). One of these is Eq. 2.2. Another can be obtained from Eq. 5.16 as the ratio of the peak amplitude to the amplitude for $\omega_p = 0$. The peak amplitude occurs at the *resonant frequency*, sometimes called the *centre frequency*, of the system. This is illustrated schematically by the *resonance curve* for driven damped simple harmonic motion shown in Fig. 5.6. The amplitude, normalised by the amplitude for $\omega_p = 0$, is plotted against the driving frequency, f_p.

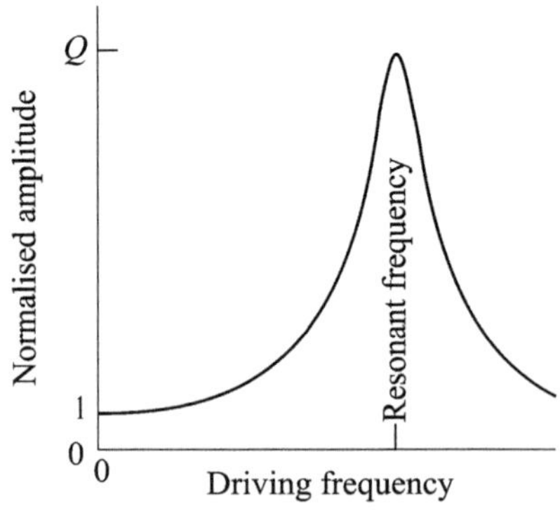

Fig. 5.6 Resonance curve for driven damped simple harmonic motion

A similar resonance curve is produced if damped simple harmonic motion is driven by a narrow band random process (Fig. 5.2). Driving damped simple harmonic motion by a broad band random process (Fig. 5.1a) results in a narrow band random process (Papoulis 1965; Pook 1984; Bendat and Piersol 2000).

5.3.2 *Random Driving*

In general, *random driving* of *damped simple harmonic motion* by a random process which is a force results in a *narrow band random process* (Fig. 5.2). For this to occur the resonant frequency (Fig. 5.6) of the damped simple harmonic motion must be within the range of frequencies for which the spectral density function of the random process shows that there is significant energy available. In other words, the spectral density function and resonance curve (Fig. 5.6) must have a significant overlap.

The spectral density function for a broad band random process (Fig. 5.1b) shows that there is significant energy is available from about 5 to 65 Hz. Using the broad band random process to drive a damped simple harmonic motion within this range of frequencies would result in a narrow band random process. In other words, a resonance is excited. For a narrow band random process the range of frequencies over which a resonance is exited is much more limited. In Fig. 5.5 it is about 0.08–0.17 Hz.

5.4 Horizontal Driving of Pendulums

The qualitative analyses below are all for small amplitudes, that is small values of the pendulum angle, θ, (Fig. 5.7). behaviour for large amplitudes is more complicated.

5.4.1 *Periodic Driving*

Consider a *simple rod pendulum* (Fig. 2.1a), whose natural period for small displacements is given by Eq. 2.13, with *horizontal driving*, as an imposed periodic displacement, applied perpendicular to the axis of the frictionless pivot. This is an example of *periodic driving*. If the imposed displacement is a sine wave (Fig. 2.5), and the pendulum angle, θ, (Fig. 5.7) is always small, then there are two possible modes of oscillation (Lamb 1923). If the period of the imposed displacement is greater than the natural period of a simple rod pendulum, rod length, l, then the motion of the point mass, m, and the imposed displacement are in phase, and the pendulum oscillates about a *virtual frictionless pivot* which is above the actual frictionless pivot (Fig. 5.7a). If the period of the imposed displacement is less than

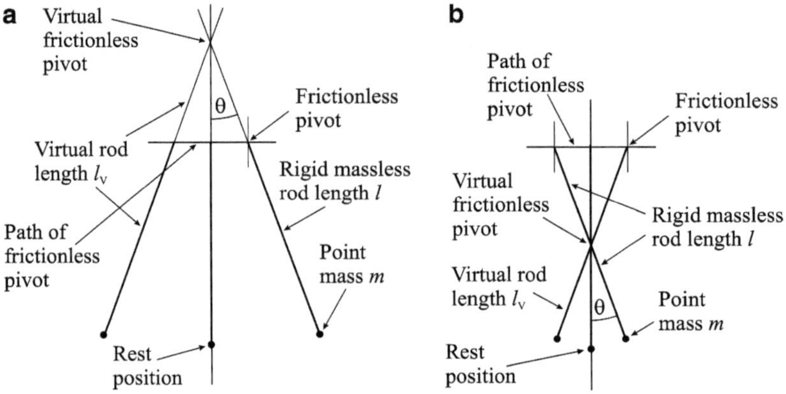

Fig. 5.7 Horizontal forcing of simple rod pendulum. Pendulum angle and point mass moving in phase. (**a**) Virtual frictionless pivot above frictionless pivot. (**b**) Virtual frictionless pivot below frictionless pivot. Pendulum angle and point mass moving out of phase

that of a simple rod pendulum, rod length, l, then the motion of the point mass, m, and the imposed displacement are out of phase, and the pendulum oscillates about a virtual frictionless pivot which is below the actual frictionless pivot (Fig. 5.7b).

For both modes of oscillation the approximate *forced period* of the pendulum, T, is given by

$$T = 2\pi\sqrt{\frac{l_V}{g}} \tag{5.18}$$

which is Equation 2.13 with the rod length, l, replaced by the virtual rod length, l_V. The two modes of oscillation are similar to those of the lower rod of a double rod pendulum, shown in Fig. 3.3a, b. If the driving period is equal to the natural period then the virtual rod length, l_V, is indeterminate, and pendulum behaviour is chaotic. The energy introduced into the system means that the amplitude of the simple rod pendulum increases with time, and the assumption that the pendulum angle, θ, is always small is eventually violated.

A *damped simple rod pendulum* is a simple rod pendulum in which a resistance torque, opposed to the motion of the pendulum, is proportional to the angular velocity of the pendulum (Sect. 4.2.2). If the horizontal driving, described above, is applied to a damped simple rod pendulum then, by analogy with driven damped simple harmonic motion (Sect. 5.3.1), the amplitude of a *driven damped simple rod pendulum* reaches a stable state where the energy introduced just equals the energy absorbed. Depending on conditions, the assumption that the pendulum angle, θ, is always small might not be violated. For small amplitudes chaotic behaviour occurs when the driving period is approximately or exactly equal to the natural period.

The above analysis for small amplitudes also applies to a *simple string pendulum* (Fig. 2.1b), driven at the clamp, provided that pendulum motion remains confined to a vertical plane containing the direction of the horizontal driving. In this special case the modes of oscillation shown in Fig. 5.7 are also possible for a simple string pendulum.

The behaviour of a real string pendulum, driven at the clamp, is more complicated. For convenience call motion in a plane containing the horizontal driving *left-right motion* and motion perpendicular to this *front-to-back motion*. When the driving period is approximately equal to the natural period of the pendulum behaviour becomes chaotic and inevitable imperfections mean that front-to-back motion takes place, as well as left-right motion (Tritton 1986, 1992). Left-right motion oscillations are at the forced period of the pendulum, and front-to back motions are at the natural period of the pendulum. In general, the combination of the two motions results in an elliptical motion which precesses (Sect. 2.4). The slightly different periods mean that the phase between the two oscillations changes so the ellipses change shape and, at times, become circles or arcs.

5.4.2 *Random Driving*

Consider a damped *simple rod pendulum* (Fig. 2.1a), whose natural period for small displacements is given by Eq. 2.13, with *horizontal driving*, as an imposed random displacement applied perpendicular to the axis of the frictionless pivot. This *driven damped simple rod pendulum* is an example of *random driving*. By analogy with random driving of damped simple harmonic motion (Sect. 5.3.2) this results in an approximately *narrow band random process* (Fig. 5.2), provided that there is an appropriate overlap between the spectral density function of the random process and the resonance curve of the *damped simple rod pendulum* (Fig. 5.6). The amplitude of a narrow band random process varies with time, and the amplitude occasionally becomes relatively large. As for periodic driving (previous section) the analysis for small amplitudes also applies to a *simple string pendulum* (Fig. 2.1b), driven at the clamp, provided that pendulum motion remains confined to a vertical plane containing the direction of the horizontal driving.

5.5 Occult Uses of Pendulums

Some authors, for example Macbeth (1943) and Jurriaanse (1987) describe *occult uses of pendulums*. The pendulums used are hand held real versions of *simple string pendulums* such as that shown in Fig. 2.3. The bob of the pendulum shown in the figure was supplied, together with Jurriaanse (1987), in what is called 'The practical pendulum pack'. Inevitable slight movements of the hand result in *random driving*. The resulting chaotic motion is similar to that observed in a real version of a simple string pendulum under periodic random driving (Sect. 5.4.1), but with the additional

Fig. 5.8 Pendulum chart

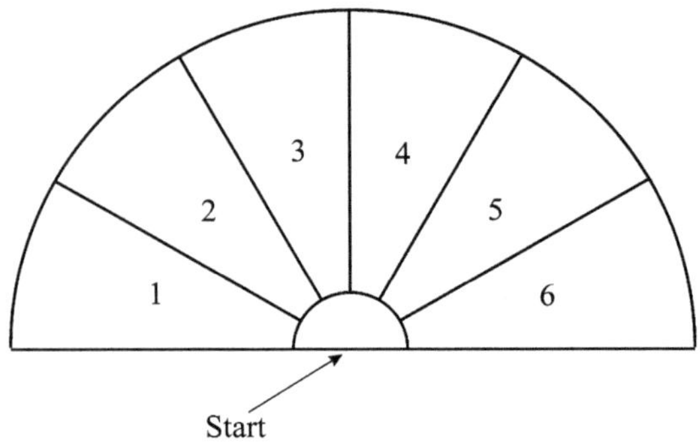

feature that the amplitude varies with time. This chaotic behaviour in hand held pendulums is described by Macbeth (1943) and Jurriaanse (1987) who ascribe it to *occult influences*.

Macbeth (1943) states that 'You may have heard of some people finding that a gold ring hung from a thread or a strand of hair, and then held in the middle of a wine glass, rings against the glass quite automatically.'. He then goes on to ascribe this to an occult influence whereas it is simply a consequence of the chaotic behaviour of a hand held pendulum. The random driving results in a two dimensional *narrow band random process*. The amplitude varies with time and, in accordance with appropriately generalised versions of Eqs. 5.13 and 5.14, eventually reaches a large enough value for the ring to strike the glass.

Jurriaanse (1987) includes a number of *pendulum charts*, intended for *occult predictions*. A pendulum chart is semi circular, and divided into several sectors, as shown schematically in Fig. 5.8. Actual charts have named, rather than numbered, sectors, and are intended to answer questions such as 'What is my favourite colour?' and 'what herb should be used to treat a headache?'. In use, a pendulum is held over the start position. Its chaotic behaviour, similar to that described in Sect. 5.4.1, ensures that it will eventually oscillate over, and hence select, one of the sectors. One might just as well throw a die (Chap. 1).

References

Baker GL, Gollub JP (1996) Chaotic dynamics. An introduction, 2nd edn. Cambridge University Press, Cambridge

Bateman D (1977) Vibration theory and clocks. Part 2. Forced harmonic motion. Horological J 120(2):48–52

Bendat JS, Piersol AG (2000) Random data: analysis and measurement procedures, 3rd edn. Wiley, New York

Bishop SR, Xu DL, Clifford J (1996) Flexible control of the parametrically excited pendulum. Proc R Soc Lond 452:1789–1806

Butikov EI (2008) Extraordinary oscillations of an ordinary forced pendulum. Eur J Phys 29(2):215–233

D'Humieres D, Beasey MR, Huberman BA, Libchaber A (1982) Chaotic pendulum states and routes to chaos in the forced pendulum. Phys Rev A 26(6):3483–3496

Frost NE, Marsh KJ, Pook LP (1974) Metal fatigue. Clarendon, Oxford

Jurriaanse D (1987) The practical pendulum book, with instructions for use and thirty-eight pendulum charts. The Aquarium Press, Wellingborough

Lamb H (1923) Dynamics, 2nd edn. Cambridge University Press, Cambridge

Macbeth N (1943) About pendulums. Pendulum play. A scientific pastime needing no knowledge of science etc. Michael Houghton, London

Papoulis A (1965) Probability, random variables, and stochastic processes. McGraw-Hill Book Company, New York

Pippard AB (1988) The parametrically maintained Foucault pendulum and its perturbations. Proc R Soc Lond A 420:81–91

Pook LP (1978) An approach to practical load histories for fatigue testing relevant to offshore structures. J Soc Env Eng 17–1(76): 22–23, 25–28, 31–35

Pook LP (1984) Approximation of two parameter Weibull distributions by Rayleigh distributions for fatigue testing. NEL Report 694. National Engineering Laboratory, East Kilbride

Pook LP (1987) Random load fatigue and r. m. s. NEL Report 711. National Engineering Laboratory, East Kilbride

Pook LP (1989) Spectral density functions and the development of Wave Action Standard History (WASH) load histories. Int J Fatigue 11(4):221–232

Pook LP (2007) Metal Fatigue: What it is, why it matters. Springer, Dordrecht

Rõssler OE, Stewart HB, Wiesenfeld K (1990) Unfolding a chaotic bifurcation. Proc R Soc Lond 431(1882):371–383

Tritton DJ (1986) Ordered and chaotic motion of a forced spherical pendulum. Eur J Phys 7(3):162–169

Tritton DJ (1992) Chaos in the swing of a pendulum. In: Hall N (ed) The new scientist guide to chaos. Penguin, London, pp 22–32

Chapter 6
Scientific Instruments

6.1 Introduction

Pendulums are used in various scientific instruments, either to make measurements, or to demonstrate scientific principles. Four of these uses are described in this chapter. These are Kater's pendulum, Newton's cradle, Foucault's pendulum, and the Charpy impact testing machine.

If the effective length, l, of a pendulum and the pendulum period, P, are known then the value of the acceleration due to gravity, g, can be calculated by using Eq. 2.15. The practical difficulty is the measurement of the effective length of a real pendulum. At the suggestion of F W Bessel in 1817 Captain Henry Kater of the British Army developed a reversible pendulum, known as *Kater's pendulum*, whose effective length can be accurately determined (Baker and Blackburn 2005). The general arrangement of a Kater's pendulum is shown in Fig. 6.1. Kater's pendulum, and subsequent refinements, were used for the determination of g until they were superseded by free-fall gravimeters in the 1950s (Torge 1980).

In its usual form, a *Newton's cradle* (Fig. 6.2a), consists of five identical pendulums each with a hardened steel ball suspended by a pair of strings with the balls just in contact. In the figure the vertical length of each pendulum, to the centre of the balls, is 18 cm. If the ball at one end of the row is pulled back and released the impulse as it hits the next ball is transmitted along the row, and the ball at the other end flies off with the remaining balls remaining almost stationary. The Newton's cradle, sometimes called a *cradle pendulum*, and in French called le pendule de Newton (Newton's pendulum), has a long history. In 1662 Huygens pointed out that an explanation of its behaviour required conservation of both *momentum* and *kinetic energy*, although he did not use the latter term (Hutzler et al. 2004). Newton's cradle is still used to demonstrate this point (Matthews 2000; Baker and Blackburn 2005). The term is used because the foundations of mechanics, including conservation of momentum and kinetic energy were established by Isaac Newton in his Principia of 1687 (Baker and Blackburn 2005). Newton's cradle is also an amusing toy; Fig. 6.2b shows a large version on display in Swiss shopping centre.

L.P. Pook, *Understanding Pendulums: A Brief Introduction*, History of Mechanism and Machine Science 12, DOI 10.1007/978-94-007-1415-1_6,

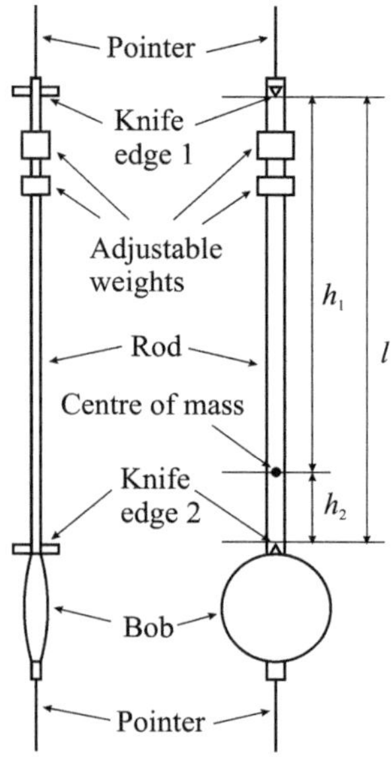

Fig. 6.1 General arrangement of Kater's pendulum

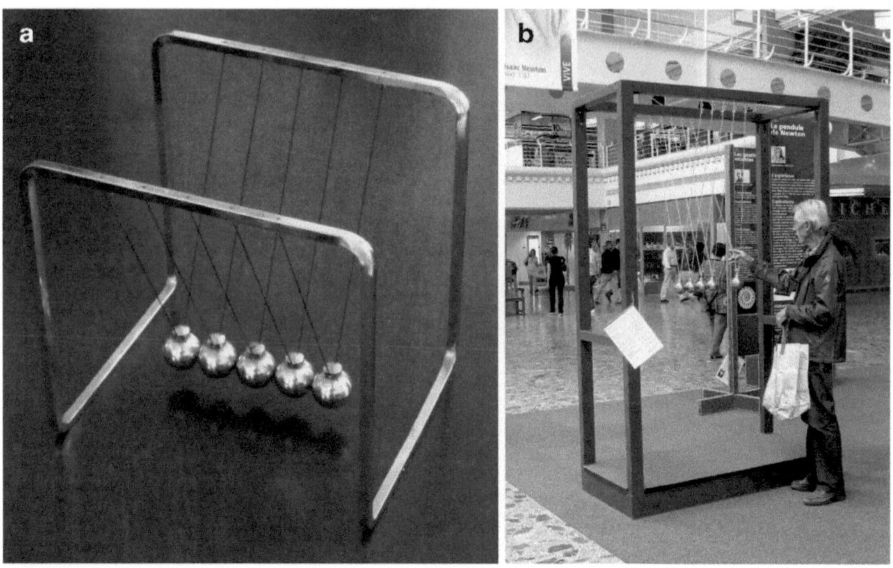

Fig. 6.2 (**a**) Typical Newton's cradle. Pendulum length 18 cm. (**b**) A large Newton's cradle in a Swiss shopping centre

The *Foucault pendulum*, sometimes called *Foucault's pendulum*, was used by Léon Foucault to demonstrate the Earth's rotation in 1851 (Matthews 2000; Tobin 2003; Baker and Blackburn 2005). Apart from their use in clocks the Foucault pendulum is probably the best known use of pendulums. There are examples in

Fig. 6.3 Foucault pendulum on display in a Swiss shopping centre

many museums around the World. A Foucault pendulum usually consists of a spherical steel bob suspended from a long steel wire, with a pointer mounted below the bob so that its motion can be easily followed. A Foucault pendulum can swing freely in any direction, and is a close approximation to a simple string pendulum (Sect. 2.4). Figure 6.3 shows a Foucault pendulum on display in a Swiss shopping centre.

In a *Charpy impact testing machine* a swinging pendulum is used to fracture notched steel test pieces. The energy absorbed in breaking a notched test piece is a useful measure of a steel's *fracture toughness*. Interest in the effects of impacts on metals dates back to the early nineteenth century (Siewert et al. 2000; Tóth et al. 2002). Impact testing of metals developed from the observation that metals are often more brittle under an impact than when loaded slowly (Siewert et al. 2000). The earliest description of an impact test appears to be that given by Tredgold (1822). He states: 'The best and most certain test of the quality of a piece of cast iron, is to try it with a hammer; if the blows of the hammer make a slight impression, denoting some degree of malleability, the iron is of good quality, providing it be uniform; if fragments off, and no sensible indentation be made, the iron will be hard and brittle.'

The first use of a pendulum impact testing machine was by S Bent Russell in 1898 (Tóth et al. 2002), and by the beginning of the twentieth century the use of pendulum impact testing machines was well established. The pendulum impact testing machine developed by Georges Charpy (Charpy 1904) is the basis of the Charpy impact testing machines that are now extensively used world wide. Methods

Fig. 6.4 Charpy impact testing machine pendulum

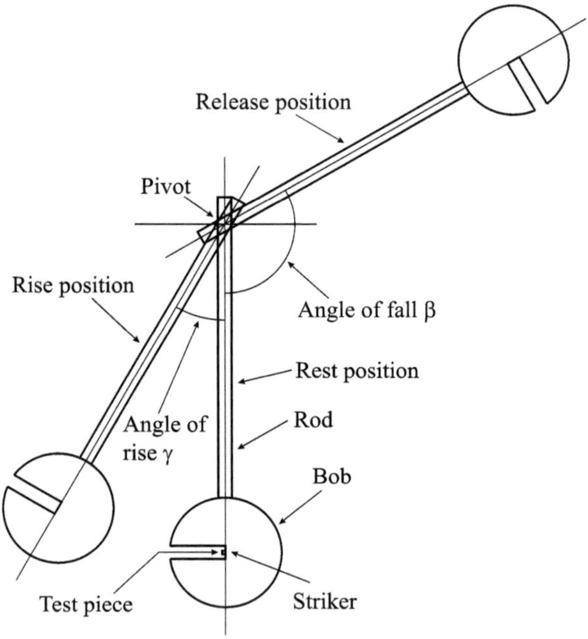

of test have been standardised, for example Anonymous (1989), although there are some variations. The general arrangement of a Charpy impact testing machine pendulum is shown in Fig. 6.4. Despites its importance the *Charpy impact test* is not usually mentioned by writers on pendulums.

6.2 Kater's Pendulum

The primary objective of Jean Richer's voyage from France to Cayenne in 1672/3 was to determine the value of solar parallax and to correct tables used by navigators and astronomers (Matthews 2000). A secondary objective was to check the reliability of pendulum clocks, which were to be used to determine the exact longitude of Cayenne. An unexpected finding was that clocks fitted with seconds pendulums ran more slowly in Cayenne (latitude 4.56° N) than they did in Paris (latitude 48.52° N). To keep correct time the pendulums had to be shortened by about 2.8 mm. The implication (Eq. 2.15) was that the acceleration due to gravity, g, was lower in Cayenne than in Paris, whereas it had previously been assumed that it was constant over the Earth's surface.

If the effective length of a real pendulum, l, and its period, P, are both known accurately, and the pendulum amplitude is small, then the value of g can be calculated using Eq. 2.15. The practical difficulty is the determination of the value of l for insertion in the equation. *Kater's pendulum* (Lamb 1923; Rawlings 1993; Baker and Blackburn 2005) is based on the finding (Sect. 3.2) that, for a compound pendulum suspended from a horizontal frictionless pivot (Fig. 3.1), the frictionless pivot and centre of oscillation are interchangeable. In other words, if the pendulum is suspended from a second frictionless pivot at the centre of oscillation corresponding to the original frictionless pivot then the pendulum period is the same. The distance between the pivots is therefore the effective length of the pendulum. Knife edge suspensions (Sect. 4.4.3), a fixed distance apart, are used in Kater's pendulum (Fig. 6.1). Kater made his knife edges as sharp as possible. However, if they have a small radius it can be shown that the effective length is still the distance between the knife edges (Rawlings 1993).

In a Kater's pendulum the knife edges are about a metre apart so the pendulum period is about 1 s. The general arrangement is shown in Fig. 6.1. The brass rod is of rectangular section with a heavy bob at the lower end and small adjustable weights near the upper end. The larger (upper) adjustable weight is clamped to the rod by a set screw and is used for coarse adjustments. The smaller weight is used for fine adjustments and is moved by a screw. There are pointers at the ends of the Kater's pendulum to facilitate observation. In use the Kater's pendulum is set up near a precision clock and the weights adjusted until the pendulum period, for a small amplitude, is the same for both knife edge suspensions. Moving a weight towards a suspension decreases the pendulum period when that suspension is used, and increases it when the other suspension is used.

To ensure that determination of the effective length is unambiguous a Kater's pendulum is deliberately asymmetric in a vertical direction. The rod pendulum, length L, shown in Fig. 3.2, is symmetrical in a vertical direction. It has the same period if suspended from either end, but the effective length is $2L/3$, not L.

Adjusting a Kater's pendulum is tedious and the observed pendulum periods are never quite equal. If P_1 and P_2 are the nearly equal pendulum periods when the Kater's pendulum is suspended from knife edges 1 and 2 (Fig. 6.1) respectively, and h_1 and h_2 are the distances of the knife edges from the centre of mass, then it can be shown that for small amplitudes the pendulum period, P, is given by (Lamb 1923)

$$P = \sqrt{\frac{P_1^2 + P_2^2}{2} + \frac{P_1^2 - P_2^2}{2}\left(\frac{h_1 + h_2}{h_1 - h_2}\right)} \tag{6.1}$$

Provided that h_1 and h_2 are distinctly different, as in Fig. 6.1, then the second term under the square root sign is very small compared with the first, and the position of the centre of mass can be found with sufficient accuracy by balancing the Kater's pendulum on a knife edge. The value of P obtained from Eq. 6.1, and the distance between the knife edges, l, which is the effective length of the pendulum, can be entered into Eq. 2.15 to determine g. As an example (Lamb 1923), let $P_1 = 0.9242$ s, $P_2 = 0.9239$ s, $h_1 + h_2 = 0.8488$ m, $h_1 - h_2 = 0.55$ m. From Eq. 6.1 $P = \sqrt{0.853868 + 0.000428} = 0.924281$ s, and from Eq. 2.15 $g = 9.806$ m/s^2.

6.3 Newton's Cradle

In its usual form, a *Newton's cradle* (Fig. 6.2) consists of five identical pendulums. Each pendulum is a real bifilar pendulum version of a bifilar pendulum (Sect. 3.5). The bob is a hardened steel ball suspended by a pair of strings. When needed for clarity this version is called a five ball Newton's cradle. In their rest positions the balls are just in contact, and this is the rest position of the Newton's cradle as a whole. The behaviour of a bifilar pendulum (Fig. 3.6), is the same as that of a simple rod pendulum (Sect. 2.3), provided that the strings remain taut. It then has one degree of freedom, and one mode of oscillation in which the point mass, m, moves in an arc in a vertical plane. This is also the primary mode of oscillation of a real bifilar pendulum. The centre of oscillation is close to the centre of mass, and the pendulum quality, Q, (Sect. 4.2.2) is high.

In the real bifilar pendulums the strings are attached at the top of the steel ball a short distance apart, so there are two additional degrees of freedom, and two secondary modes of oscillation are possible. These are an oscillation in the plane of the strings and a torsional oscillation, The three degrees of freedom, and the three modes of oscillation are analogous to those of the dual string pendulum (Sect. 3.7.1). A closer analogy for the mode of oscillation in the plane of the strings is the trapezium pendulum (Sect. 3.8). The pendulum quality for the secondary modes of oscillation is low. If minimised by careful construction and alignment of a Newton's cradle these secondary modes of oscillation do not need to be considered (Sect. 4.4). In the analysis below it is therefore assumed that motions of the individual real bifilar pendulums correspond to motions in the primary mode of oscillation.

6.3.1 Modes of Oscillation

A five ball *Newton's cradle* as a whole has five primary modes of oscillation, and these involve all five real bifilar pendulums. In these modes of oscillation interference between the real bifilar pendulums means that overall pendulum quality, Q, is low, even though the quality of individual pendulums is high, and oscillations decay rapidly. The five modes can easily be demonstrated by experimenting with a Newton's cradle, such as those shown in Fig. 6.2. The *time of swing* is approximately the same for all five modes. It is defined as the time interval for a given configuration to be repeated.

To start the first primary mode of oscillation, from the rest position, one ball at an end of the row is pulled back and released. The impulse as it hits the next ball is transmitted along the row, and the ball at the other end of the row flies off, with the other four balls remaining almost stationary. This ball then swings back and hits the row. The oscillation continues with a ball flying off each end of the row alternately. The central group of three balls remains almost stationary throughout the oscillation.

To start the second primary mode of oscillation, from the rest position, a pair of balls at an end of the row, kept in contact, is pulled back and released. The impulse as the pair of balls hits the next ball is transmitted along the row, and a pair of balls at the other end of the row flies off, still in contact, with the other three balls remaining almost stationary. The pair of balls then swings back and hits the row. The oscillation continues with a pair of balls flying off each end of the row alternately. The central ball remains almost stationary throughout the oscillation.

To start the third primary mode of oscillation, from the rest position, a group of three balls at an end of the row, kept in contact, is pulled back and released. The impulse as the group of three balls hits the next ball is transmitted along the row, and a group of three balls at the other end of the row flies off, still in contact, with the other two balls remaining almost stationary. The group of three balls then swings back and hits the row. The oscillation continues with a group of three balls flying off each end of the row alternately. The central ball swings continuously throughout the oscillation.

To start the fourth primary mode of oscillation, from the rest position, a group of four balls at an end of the row, kept in contact, is pulled back and released. The impulse as the group of four balls hits the single ball is transmitted along the row, and a group of four balls at the other end of the row flies off, still in contact, with a single ball remaining almost stationary. The group of four balls then swings back and hits the single ball. The oscillation continues with a group of four balls flying off each end of the row alternately. The central group of three balls swings continuously throughout the oscillation.

To start the fifth primary mode of oscillation all five balls, kept in contact, are pulled back and released. The five balls then swing in unison in a *uniform mode of oscillation* in which all five balls stay in contact and move in phase.

The first four primary modes of oscillation, in which balls fly off alternately from each end of the row, are called *fly off modes of oscillation*. These modes are unstable and degenerate steadily into the fifth primary mode of oscillation. The collisions between the balls means that the real bifilar pendulums are *coupled pendulums* in which energy is transferred between pendulums. Coupled pendulums, with identical or nearly identical periods, have a tendency to become synchronised, as first observed by Huygens in 1665 (Baker and Blackburn 2005). This tendency to synchronisation is the reason why the fly off modes of oscillation degenerate into the fifth primary mode of oscillation.

An N ball Newton's cradle has N primary modes of oscillation. There are $N - 1$ fly off modes of oscillation, and there is one uniform mode of oscillation. Thus, for a two ball Newton's cradle consisting of two real bifilar pendulums there are two primary modes of oscillation. The first primary mode of oscillation is a fly off mode of oscillation in which one ball is pulled back from the rest position and released. The impulse as the first ball hits the second ball is transmitted to the second ball, which flies of with the first ball remaining almost stationary. The second ball swings back and the oscillation continues. The second primary mode of oscillation is a uniform mode of oscillation in which the two real bifilar pendulums swing in unison. Experiments with N up to five shows that degeneration of fly off modes of oscillation into a uniform mode of oscillation occurs increasingly rapidly as N increases. Numerical simulations for a two ball Newton's cradle (Hutzler et al. 2004) confirm that the first primary (fly off) mode of oscillation does degenerate into the second primary (uniform) mode of oscillation, as is observed experimentally.

Secondary modes of oscillation of a Newton's cradle include various *bouncing modes of oscillation*. For example, if the balls of a two ball Newton's cradle are pulled apart by the same distance and released simultaneously the balls bounce off each other at the rest position and each ball swings in a succession of half cycles.

6.3.2 Theoretical Analysis

A five ball *Newton's cradle* is an example of a system that appears to be simple but is actually complex and difficult to analyse in detail (Baker and Blackburn 2005). The usual textbook analysis of Newton's cradles provides a reasonable approximation for the first cycle of a fly off mode of oscillation. It takes as a starting point Huygens' 1662 observation (Hutzler et al. 2004) that theoretical analysis of a Newton's cradle must be based on conservation of both *momentum* and *kinetic energy*. As simplifications it is assumed that the balls move along the same straight line in simple harmonic motion (Sect. 2.2), and that the system is non dissipative. In other words once started a Newton's cradle continues to swing indefinitely. The assumption that the system is non dissipative includes the implicit assumption that collisions between the balls are *perfectly elastic*. That is, all the kinetic energy absorbed during the elastic deformation as the balls collide is returned as kinetic energy as the balls return to their undeformed shape. In physics terminology the *coefficient of restitution* is one. The point of using hardened steel balls is that collisions are essentially elastic so negligible energy is lost during a collision.

In the usual textbook analysis, consider a two ball Newton's cradle consisting of two identical real bifilar pendulums swinging in the first primary mode of oscillation (previous section). If immediately before a collision between the two balls the velocity of the right hand ball, mass m, is zero, then its momentum and kinetic energy are also zero. If at the same time the velocity of the left hand ball, mass M, is V, then its momentum, p, is given by (Baker and Blackburn 2005)

$$p = MV \tag{6.2}$$

and its kinetic energy, K, is given by

$$K = \frac{1}{2}MV^2 \tag{6.3}$$

which is Eq. 2.20 with m replaced by M.

Immediately after the collision both momentum and kinetic energy are conserved so they are zero for the left hand ball, and are given by Eqs. 6.2 and 6.3 for the right hand ball. During the collision the left hand ball applies an impulse, over a very short period of time, to the right hand ball, and the right hand ball applies an impulse to the left hand ball. The impulses are assumed to be short enough to be regarded as instantaneous. These impulses are of the same magnitude but opposite in sign, and have a much more drastic effect than do those of a clock escapement (Sect. 4.5).

During an elastic collision some of the kinetic energy is stored as *strain energy* (Gere and Timoshenko 1991), and then returned. When, as is usual, the coefficient of restitutions is less than one, then some energy is dissipated as heat.

The usual textbook analysis is satisfactory for a two ball Newton's cradle, but does not provide a complete explanation for Newton's cradles with more than two balls. This is because conservation of both momentum and kinetic energy does not lead to a unique solution for continuous oscillations (Hutzler et al. 2004; Baker and Blackburn 2005). It is satisfactory for the first oscillation, since the fact that the number of balls pulled back from the rest position and released is always the same as the number of balls that fly off the other end of the row is indeed a simple consequence of the conservation of both momentum and kinetic energy (Matthews 2000).

6.3.2.1 Phase Diagrams

For the first oscillation, the motion of each of the real bifilar pendulums in a Newton's cradle can be represented by a phase diagram in which angular acceleration is plotted against angular velocity. In the usual textbook analysis the system is assumed to be non dissipative, so the phase diagrams in Figs. 6.5 and 6.6 are shown as closed circuits. Figure 6.5a shows the phase diagram for the left hand ball of a two ball Newton's cradle in the first primary mode of oscillation. One complete circuit of the phase diagram, in the direction shown by the arrow,

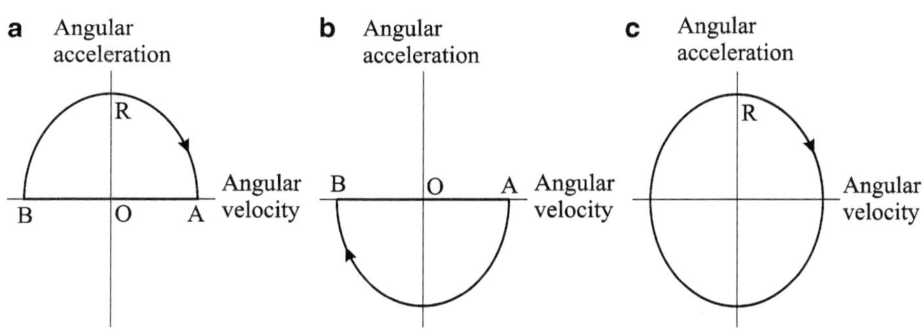

Fig. 6.5 Phase diagrams for a two ball Newton's cradle, first oscillation. *Arrows* show the sense of time. (**a**) First primary mode of oscillation, left hand ball. (**b**) First primary mode of oscillation right hand ball. (**c**) Second primary mode of oscillation, both balls

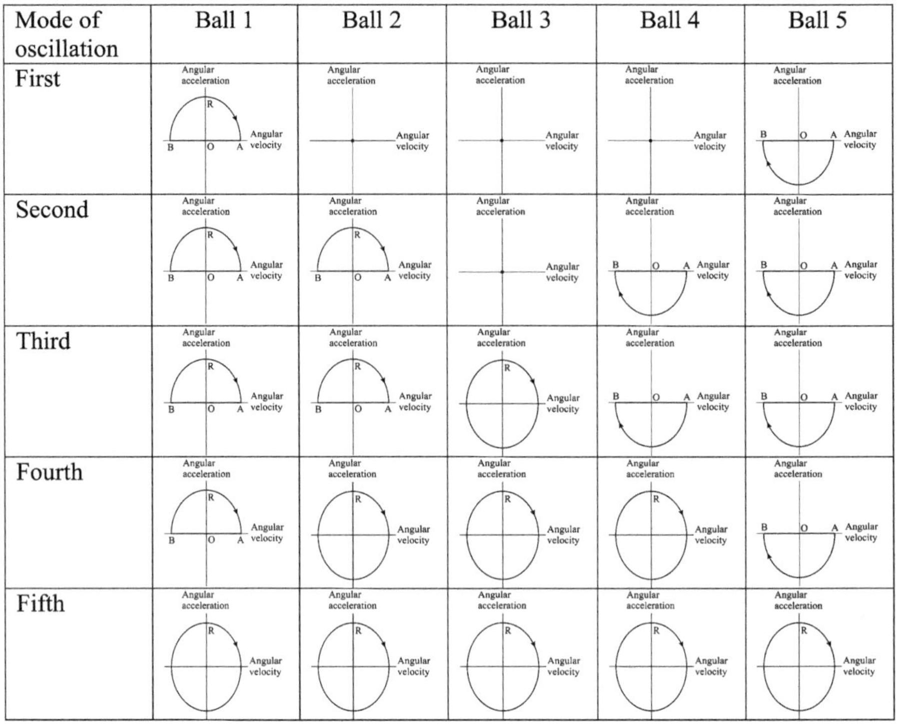

Fig. 6.6 Phase diagrams for the primary modes of oscillation of a five ball Newton's cradle, first oscillation. The balls are numbered from *left* to *right*. *Arrows* show the sense of time

represents the behaviour of the left hand ball during one complete oscillation of the Newton's cradle. If the left hand ball is pulled to the left and released it starts at point R. It then swings freely along the arc RA for one quarter of the time of swing. At point A it receives an impulse from the right hand ball, represented by the straight line AO, and remains stationary in the rest position, represented by point O, for half the time of swing. It then receives an impulse from the right hand ball, represented by the straight line OB. The ball then swings freely, represented by the arc BR for one quarter of the time of swing, completing the oscillation. The phase diagram shown in Fig. 6.5b represents the right hand ball. It is stationary in the rest position, represented by the straight line, until it receives an impulse from the left hand ball, represented by the straight line OA for one quarter of the time of swing, The ball then swings freely along the arc AB for half the time of swing. It then receives an impulse from the left hand ball, represented by the straight line BO, and remains stationary in the rest position, represented by point O for one quarter of the time of swing, completing the oscillation. The phase diagram for both balls in the second primary mode of oscillation is shown in Fig. 6.5c (cf Fig. 2.9). If both balls are pulled to the left and released simultaneously, they start at point R and swing freely round the circuit, completing the oscillation.

Phase diagrams for the first oscillation of all five primary modes of oscillation of a five ball Newton's cradle are shown as thumbnails in Fig. 6.6. The balls are numbered from left to right. There are four different types of phase diagram. For example, in the first primary mode of oscillation the phase diagrams for balls 1 and 5 are the same as those shown Fig. 6.5a, b respectively, and balls 2, 3, and 4 are, nominally, stationary so the phase diagram is a single point at the origin: it is shown as a dot in the figure. In the third primary mode of oscillation ball 3 swings continuously and the phase diagram is the same as Fig. 6.5c.

6.3.2.2 Elastic Collisions

The usual textbook analysis of Newton's cradle (Sect. 6.3.2) leads to the conclusion that, in fly off modes, there are three types of ball behaviour. These are illustrated by the phase diagrams for the first oscillation of a five ball Newton's cradle shown in Fig. 6.6. Balls at the ends of a row are stationary during half a cycle. Balls within the row can be either stationary during half a cycle, completely stationary, or swing continuously. In practice, collisions are elastic, balls can rebound, and the line of balls breaks up after the first collision. This is shown by careful observation of a Newton's cradle, and has been demonstrated by numerical simulation (Hutzler et al. 2004). Rebounds mean that, contrary to the usual textbook analysis, all the balls move so do not remain stationary during half cycles.

Hutzler et al's simulations show that, for a non dissipative Newton's cradle, behaviour in fly off modes of oscillation changes from cycle to cycle, and that fly off modes of oscillation do not degenerate into uniform modes. Hence, a non dissipative Newton's cradle is always a chaotic system. Chaotic system are sensitive to initial conditions. Slight differences in how the system is set into motion lead to significant differences in subsequent behaviour, which is both irregular and apparently unpredictable

Some of Hutzler et al.'s results are summarised qualitatively in the phase diagrams for the first primary mode of oscillation of a five ball Newton's cradle shown in Fig. 6.7. The balls are numbered from left to right. Ball 1 is pulled back and released, and swings along the arc RA (Fig. 6.7a). The impulse as it collides with ball 2 passes along the line of balls. Ball I rebounds and moves along the arc CD, ball 2 moves along the arc CD (Fig. 6.7b), ball 3 only moves slightly so its phase diagram is represented by a dot (Fig. 6.7c), ball 4 moves the arc BA (Fig. 6.7d), ball 5 flies off the end of the row and moves along the arc AB (Fig. 6.7e). The impulse as it flies back and collides with ball 4 passes back along the line of balls. Ball 5 rebounds and passes along the arc DC (Fig. 6.7e), ball 4 moves along the arc CD (Fig. 6.7d), ball 3 only moves slightly so its phase diagram is represented by a dot (Fig. 6.7c), ball 2 moves along the arc AB (Fig. 6.7b), ball 1 flies off and moves along the arc BR completing the first cycle. The second cycle is similar. The phase diagram for ball 3 on the first cycle is similar to Fig. 6.7a, but on a much smaller scale. On the second cycle it is similar to Fig. 6.7d, again on a much smaller scale.

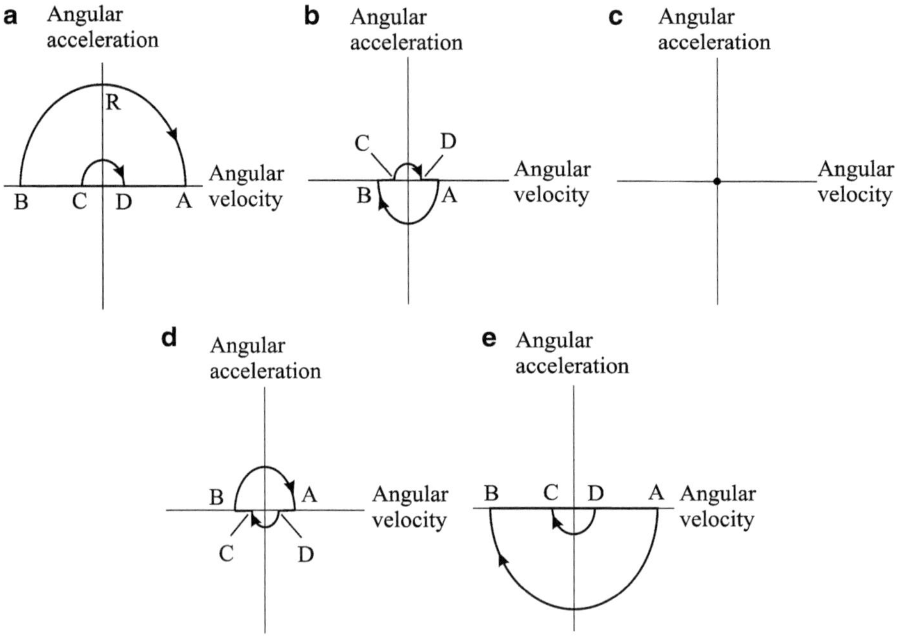

Fig. 6.7 Phase diagrams for a five ball Newton's cradle including effect of rebounds. First primary mode of oscillation. The balls are numbered from *left* to *right*. (**a**) Ball 1. (**b**) Ball 2. (**c**) Ball 3. (**d**) Ball 4. (**e**) Ball 5

6.3.2.3 Energy Losses

In practice a Newton's cradle is a dissipative system, so the oscillations decay, and the phase diagrams spiral inwards, as in Fig. 4.6. The main source of *energy loss* in a carefully aligned Newton's cradle appears to be *friction* between balls. In the usual textbook analysis of Newton's cradle (Sect. 6.3.2) it is assumed that the balls move in a straight line. In fact the balls move along different arcs so that there is relative up and down motion, and concomitant friction, between adjacent balls that are in contact. It is this friction that appears to be the cause of the poor pendulum quality of a Newton's cradle, as compared to the quality of the individual real bifilar pendulums. (Sects. 6.3 and 6.3.1). The lost energy appears as heat.

Hutzler et al. (2004) assume that energy loss in a Newton's cradle is due to the viscoelastic dissipation associated with *collisions* of the balls and velocity dependent *drag* of air. In general, there is some loss of energy during a collision. The usual empirical approach, which dates back to Newton (Lamb 1923), is that the velocity of say, a bouncing ball, is reversed before and after the collision is reversed, and that the *coefficient of restitution* is the arithmetic ratio of the velocity of the ball immediately after impact to that immediately before impact. This is what happens in *bouncing modes of oscillation* (Sect. 6.3.1). Hence the coefficient of restitution never exceeds one, and in practice is always less than one since some energy is dissipated as heat (Sect. 6.3.2). The use of hardened steel balls minimises the energy lost during collisions (Sect. 6.3.2), and the energy lost appears to be small compared with that due to friction. Data presented by Matthys (2004) suggest that the energy lost by *air drag* is relatively small.

6.4 The Foucault Pendulum

A *Foucault pendulum*, as used by Léon Foucault to demonstrate the Earth's rotation in 1851, is not a distinct type of pendulum. It is a large real version of a simple string pendulum (Sect. 2.4), and therefore has two degrees of freedom. Foucault's original pendulum was a 5 kg ball swinging on a 2 m wire, he then tried an 11 m wire and finally a 28 kg ball on a 67 m wire in the Panthéon in Paris (Matthews 2000; Tobin 2003). Figure 6.3 shows a Foucault pendulum on display in a Swiss shopping centre.

The appearance of a physical system, such as a Foucault pendulum, to an observer depends upon what *frame of reference* is used (Baker and Blackburn 2005). The physical system is unaffected by the choice of frame of reference. For example, a passenger sitting in an aeroplane in flight is moving at the speed of the aeroplane if the Earth's surface is used as the frame of reference, but is stationary if the aeroplane is used as the frame of reference.

However, the Earth rotates relative to the sun and the distant stars, which are regarded as fixed in space, and is a *rotating frame of reference*. It is this rotation that can be demonstrated by using a Foucault pendulum. By definition, the mean time for one rotation of the Earth, T, relative to the sun, that is 1 day, is 24 h. The *sideral day*, defined by rotation relative to distant stars (Matthews 2000; Tobin 2003) is 23.9344696 h (23 h 56 m 4.1 s). Equations of motion of a Foucault pendulum in a rotating frame of reference are complex (Barnard 1874; Synge and Griffith 1959; Baker and Blackburn 2005). Closed form solutions exist for small amplitude motions (Condurache and Martinual 2008).

A Foucault pendulum that is suspended from a point that is fixed in space, and set swinging in a *planar mode of oscillation* so that the path of the point mass is an arc in a vertical plane, continues to swing in the same arc when viewed by an observer who is in a frame of reference that is fixed in space. However, the path of the bob of a Foucault pendulum suspended from a point that is fixed relative to the Earth, and set swinging in an arc in a vertical plane does not continue to swing in the same arc, but follows a more complicated path to an observer who is in a frame of reference that is fixed to the Earth's surface. The deviations from a planar mode of oscillation are relatively small.

In general, when considering pendulums, the Earth may be regarded as a fixed frame of reference. In particular, the design of clock pendulum *suspensions* (Sect. 4.4) ensures that deviations from a planar mode of oscillation are suppressed.

By selection of an appropriate frame of reference, a Foucault pendulum suspended from a point that is fixed relative to the Earth's surface can swing in a planar mode of oscillation. The frame of reference rotates about an axis that is perpendicular to the Earth's surface and passes through the point of suspension. If the time of rotation of the Earth is T then, since rotation is a vector, the time of one rotation of the frame of reference, T_E, is given by (Barnard 1874; Synge and Griffith 1959; Matthews 2000; Tobin 2003; Baker and Blackburn 2005; Phillips 2005)

$$T_E = \frac{T}{\sin L} \tag{6.4}$$

where L is latitude. T_E can be interpreted to be the effective time of one rotation of the Earth, as given by rotation of the plane in which a Foucault pendulum swings.

To an observer who is in a frame of reference that is fixed to the Earth's surface the path of the bob of a Foucault pendulum is a combination of oscillations along an arc and rotation about the vertical axis which passes through the midpoint of the arc (Condurache and Martinual 2008). It is this precession that demonstrates rotation of the Earth. Looking downwards rotation is clockwise in the Northern Hemisphere and anticlockwise in the southern Hemisphere. At the North and South Poles $T_E = T$, and at the Equator $T_E \to \infty$. The latitude of London is 51.30° N, so $T_E = 1.281$T. For a pendulum use of the sideral day is appropriate, so T_E for London is 30 h 40 m 6 s.

For small amplitudes oscillations along the arc approximate to simple harmonic motion (Sect. 2.2). Forces on the bob are analysed by Phillips (2005). As timepieces Foucault pendulums are poor (Pippard 1988; Tobin 2003) and records are not normally kept. A Foucault pendulum at the South Pole had a period of one rotation of 24 h $\pm$ 50 m (Baker and Blackburn 2005). The precession rate of a Foucault pendulum in London, averaged over several weeks, was within 1% of the expected value (Tobin and Pippard 1994).

6.4.1 Essential Features

A *Foucault pendulum* used to demonstrate the Earth's rotation has four essential features. Firstly, it is spherical pendulum (Sect. 2.4) designed to swing equally freely in all directions. In other words the pendulum quality, Q, (Sect. 4.2.2) should be the same for all directions. Secondly, it must be able to be launched so that it swings correctly. Thirdly, once launched it must continue to swing correctly for several hours so that the Earth's rotation can become clearly apparent. Finally, it has an *indicating device* or a *recording device* to show the amount of the Earth's rotation. How to make successful Foucault pendulums of various lengths is now well understood (Betrisey 2010).

A Foucault pendulum usually consists of a spherical steel ball suspended by a steel wire (Matthews 2000; Tobin 2003; Baker and Blackburn 2005). The usual indicating device is a pointer is mounted below the bob. The amplitude is small so the pendulum period is given approximately by Eq. 2.15. The pendulum quality, Q, is fairly high. From graphical decay data (Baker and Blackburn 2005) the pendulum quality, Q, for the 67 m Foucault pendulum in the Panthéon fitted with a 47 kg bob is about 830.

A real version of a spherical rod pendulum (Sect. 2.5) was used as a Foucault pendulum by H Kamerlingh Onnes, and described in his 1879 dissertation (Schultz-DuBois 1970). The universal joint (Fig. 2.14) was a gimbal consisting of two crossed knife edge suspensions. A single knife edge suspension is shown in Fig. 4.8. This was probably the first successful short Foucault pendulum. The pendulum length appears to be in the range 1–2 m (Baker and Blackburn 2005). Construction of Foucault pendulums of this order of length is now routine (Betrisey 2010).

The standard method of launching a Foucault pendulum is the *burnt string method* (Tobin; 2003; Baker and Blackburn 2005). The pendulum bob is tethered to a fixed object by a string, and the string is then burned through.

There is usually a circular scale, marked in degrees and sometimes in hours, below the bob so that the pointer below the bob shows how much the Earth has turned since the Foucault pendulum was launched (Tobin 2003; Baker and Blackburn 2005). Sometimes sand is spread on the floor as an recording device, and is progressively marked by the pointer as it passes. Another recording device is a ring of wooden blocks that are progressively knocked over (Fig. 6.3).

6.4.2 Pumping

There are two approaches to ensuring that a *Foucault pendulum* continues to swing for several hours. In the first, a Foucault pendulum is left to swing freely after it is launched so pendulum quality is important. The pendulum period increases as the length of a pendulum increases so, for constant pendulum quality (Sect. 4.2.2), the time taken for the amplitude to decay to, say, half its initial amplitude increases with the length of the pendulum. This is one of the reasons why Foucault, in his original experiments (Sect. 6.4) used pendulums of increasing length. A freely swinging Foucault pendulum does provide convincing proof that the Earth does rotate.

In the second approach, appropriately timed impulses, sometimes called *pumping*, are applied to the pendulum to keep its amplitude constant, in much the same way that escapements keep the amplitude of a clock pendulum constant (Sect. 4.5). The drive system should be designed to interfere as little as possible with the motion of the pendulum. A pumped Foucault pendulum is less convincing than a freely swinging Foucault pendulum and should be regarded as demonstrating, rather than proving, that the Earth does rotate.

Two methods of pumping are used (Tobin 2003; Baker and Blackburn 2005). In *vertical pumping* the suspension point is pumped in a vertical direction. This can be done either mechanically or electromagnetically. The suspension point is raised as the pendulum passes through the rest position and lowered at the apses. Hence, radial pumping is at twice the frequency of the pendulum. There are four impulses for each to and fro swing of the pendulum as shown in the phase diagram

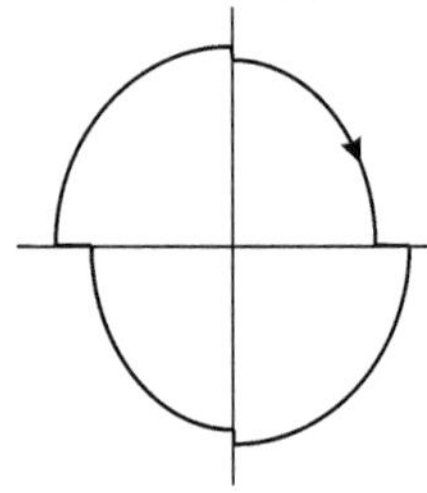

Fig. 6.8 Phase diagram for vertical pumping of a Foucault pendulum. The *arrow* shows the sense of time

(Fig. 6.8). Those at the rest position are positive, and those at the apses are negative. Because of the effects of centripetal force on the vertical component of the reaction at the suspension, the positive impulses are arithmetically greater than the negative impulses so there is a met input of energy into the pendulum. A *parametrically maintained pendulum* is driven by an external oscillator which raises and lowers the suspension point, usually sinusoidally. The frequency of the oscillation is adjusted so that the pendulum becomes unstable, and oscillates (Pippard 1988; Zevin and Filonenko 2007).

In *angular pumping* of a Foucault pendulum the bob is impulsed electromagnetically with impulses in the direction of travel in much the same way that clock pendulums are driven by an escapement (Sects. 4.1 and 4.5). Timing of the impulses is controlled by the oscillations of the pendulum. Electromagnetic radial pumping was first used by Léon Foucault in 1855 (Baker and Blackburn 2005). With modern electronics no moving parts are needed. The Foucault pendulum shown in Fig. 6.3 is pumped by an electromagnetic system, with no visible parts.

A *Charron ring*, introduced in 1931, is sometimes used to ensure that a pumped Foucault pendulum swings correctly in the desired *planar mode of oscillation* within the appropriate frame of reference (Tobin 2003; Baker and Blackburn 2005; Betrisey 2010). An accurately centred horizontal ring is set just below the suspension point of a Foucault pendulum that consists of a spherical steel ball suspended by a steel wire. The amplitude and Charron ring diameter are chosen so that the steel wire just touches the ring in the vicinity of the apses. The resulting friction helps to keep the pendulum swinging correctly, but reduces pendulum quality. The Charron ring is so effective that the pendulum can be launched by hand: the burnt string method is unnecessary.

6.4.3 Motions of the Bob

When swinging correctly, viewed from the Earth's surface the motion of the bob of a *Foucault pendulum* is an orbit, which is a combination of a rotation about an axis perpendicular to the suspension point and an oscillation along a circular arc (Phillips 2005). For the small amplitudes typical of a Foucault pendulum the latter

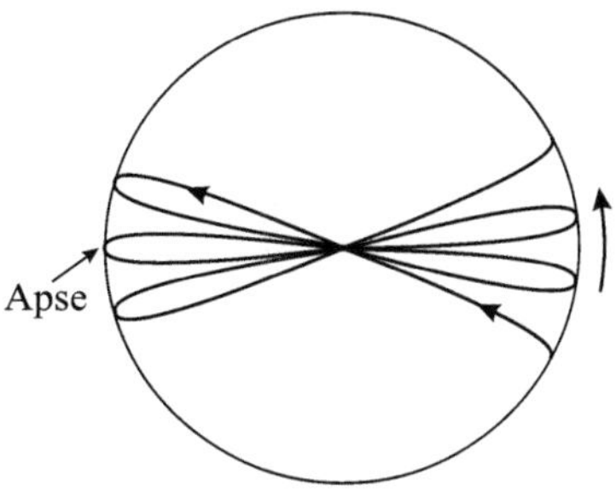

Fig. 6.9 Three complete oscillations of a point mass moving in simple harmonic motion combined with a rotation. *Arrows* show the directions of the rotation and of the motion of the point mass

can be approximated by simple harmonic motion of a point mass (Sect. 2.2). This combination is shown for three complete oscillations (apse to apse) in Fig. 6.9.

The orbit of the point mass does not repeat exactly but precesses about a vertical axis. The apses lie on a circle. For a Foucault pendulum the apses are much closer together than in the figure: there are thousands of oscillations of the bob in one complete rotation of the pendulum. At an apse the velocity of the point mass is zero in a radial direction, but it is moving in a circumferential direction. Before it is launched by the burnt string method (previous section) the bob of a Foucault pendulum is moving in a circumferential direction due to the Earth's rotation. When it is launched the bob moves in a curve towards the rest position.

A Foucault pendulum is a real version of a simple string pendulum (Sect. 2.4), and can swing in a path that, for small amplitudes, is approximately elliptical. Inevitable imperfections and perturbations mean that the path of the bob of a Foucault pendulum tends to degenerate into an ellipse. This is called *ellipsing* (Tobin 2003; Baker and Blackburn 2005; Betrisey 2010). An ellipsing pendulum precesses as shown schematically in Fig. 2.13 with the rate of precession exaggerated. Corresponding sketches in Tobin and Pippard (1994), Tobin (2003), Baker and Blackburn (2005) and Condurache and Martinual (2008) are similar. The angular interval, γ, between apses is (Synge and Griffith 1959; Tobin and Pippard 1994)

$$\gamma = \frac{3\pi ab}{4l^2} \tag{6.5}$$

where a and b are the semi major and minor axes of the ellipses and l is the effective length of the pendulum. Hence, an ellipsing Foucault pendulum precesses at an incorrect rate. Assuming that the semi major axis of the ellipses is proportional to pendulum length, and also that the ellipses are of constant ellipticity, then the precession rate in terms of time is proportional to $1/l^{1.5}$, which shows a benefit of increasing the length of a Foucault pendulum.

The elliptical mode of oscillation can be resolved, along the major and minor axes of the ellipse, into two planar modes of oscillation (Sect. 2.4). These have the same time of swing, but different amplitudes, and are 90° out of phase. This phase difference means that correctly timed pumping opposes ellipsing since oscillation along the minor axis of the ellipse is opposed (Pippard 1988). It is not clear to what extent this benefit is obtained in practice.

6.5 Charpy Impact Testing Machine

In a *Charpy impact testing machine*, shown schematically in Fig. 6.4, a swinging pendulum is used to fracture notched steel test pieces. In the current British Standard form of the *Charpy impact test* (Anonymous 1989) the test pieces are 10 mm × 10 mm × 55 mm with a V-notch on one face. The V-notch is 2 mm deep with an included angle of 45°, and the notch root radius is 0.25 mm. Other types of notch are sometimes used. A swinging pendulum (Fig. 6.4) is used to fracture the test pieces. To carry out a test, the test piece is placed on supports with the notched face against anvils 40 mm apart (Fig. 6.10). It is then loaded in bending by an impact on the face opposite the notch by a *striker* attached to a pendulum bob. The striker has a tip radius of either 2 mm or 8 mm. The energy absorbed in breaking a notched test piece is a useful measure of its *fracture toughness* or resistance to *brittle fracture*. This energy is obtained from the difference in height at the start of the pendulum swing and the height the pendulum reaches after breaking the test piece. Examination of the broken test pieces provides useful additional information (Biggs 1960). Tests on a particular steel are usually carried out over a range of temperatures.

To carry out a Charpy impact test the pendulum is raised to the release position where it is held by a catch. In the release position the potential energy, W, expressed in terms of the angle of fall, β, (Fig. 6.4) is (cf Eq. 2.19)

$$W = Mgl\,(1 - \cos\beta) \tag{6.6}$$

where M is the mass of the pendulum, g is acceleration due to gravity, and l is the effective pendulum length. The test piece, carefully centred, is placed on the test piece supports against the anvils (Fig. 6.10) and the pendulum is released. The energy absorbed in breaking the specimen means that the angle of rise, γ, is less than

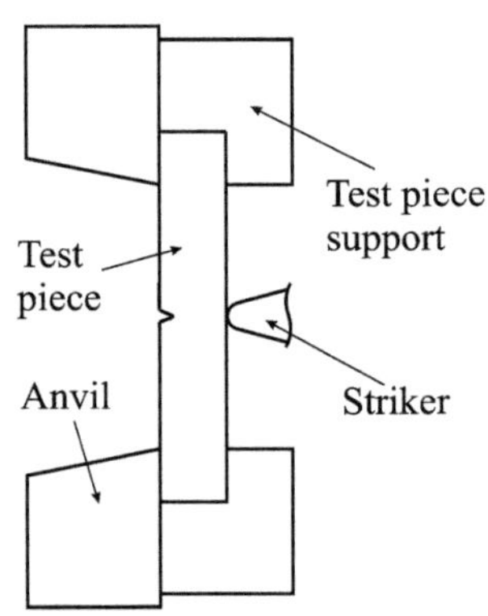

Fig. 6.10 Top view of Charpy test piece, striker, anvils and supports

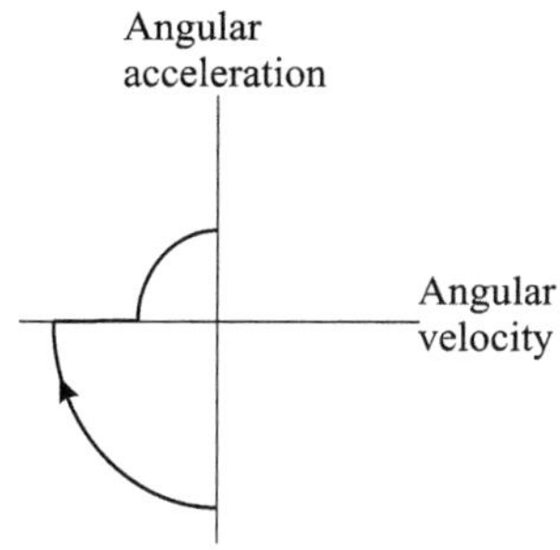

Fig. 6.11 Phase diagram for a Charpy impact testing machine pendulum. The *arrow* shows the sense of time

the angle of fall. In other words, breaking the specimen applies a large impulse to the pendulum, as shown in the phase diagram (Fig. 6.11). The absorbed energy W_a is

$$W_a = W - Mgl(1 - \cos\gamma) \tag{6.7}$$

A follower pointer, or an electronic equivalent, records the angle of rise. A scale shows the absorbed energy in joules.

The Charpy impact test is easy, quick and cheap. It is a particularly convenient quality control test for steels (Biggs 1960; Tipper 1962); but it does not provide quantitative fracture toughness data that can be used in design (Frost et al. 1974). Steel specifications sometimes include a requirements that the energy must not be below a specified energy at a specified temperature. A British Standard for Charpy impact testing was published in 1920 (Anonymous 1990a), but it was not until about 1960 that machines were sufficiently refined for results obtained to be both accurate and repeatable between different machines (Siewert et al. 2000). Current standards, for example Anonymous (1990b), include very detailed requirements for the verification of Charpy test machines, sometimes including the use of reference test pieces of known energy absorption.

A Charpy impact testing machine pendulum has a heavy bob and a light rod, so it is a close approximation of a simple rod pendulum (Sect. 2.3), and is designed to minimise energy losses. The centre of percussion (Sect. 3.2) is at the tip of the striker. Impact with the specimen is as the pendulum passes through the rest position, and the centre of percussion is within the width of the test piece. A Charpy impact testing machine is calibrated in joules. It has one or more ranges with nominal potential energies of 150 J, 300 J, etc. Calibration allows for the inevitable minor energy losses. These are difficult to assess, hence the use of reference test pieces as a check.

The current British Standard for verification of Charpy impact testing machines (Anonymous 1990b) requires that the velocity at strike be in the range 5.0 m/s to 5.5 m/s. Taking g as 9.81 m/s^2 Eq. 2.21 implies that $l(1-\cos\beta)$ must be in the range 1.274 m to 1.542 m, and hence places limits on possible combinations of pendulum length and angle of fall. Georges Charpy's pendulum drop-weight machine no. 2 (Charpy 1904) had a velocity at strike of 5.28 m/s, so its proportions are similar to those of modern Charpy impact testing machines.

References

Anonymous (1989) Notched bar tests. Part 6. Method for precision determination of Charpy V-notch impact energies for metals (BS 131: Part 6: 1989). British Standards Institution, London

Anonymous (1990a) Metallic materials – Charpy impact test – Part 1: Test method (V- and U-notches) (BS EN 10045–1: 1990). British Standards Institution, London

Anonymous (1990b) Notched bar tests. Part 6. Specification for verification of the test machine used for precision determination of Charpy V-notch impact energies for metals (BS 131: Part 7: 1990). British Standards Institution, London

Baker GL, Blackburn JA (2005) The pendulum. A case study in physics. Oxford University Press, Oxford

Barnard JG (1874) Problems of rotary motion presented by the gyroscope, the precession of the equinoxes, and the pendulum. Smithsonian Contrib Knowl 19(240):1–52

Betrisey M (2010) Works of art that also tell the time. http://www.betrisey.ch/eindex.htm Accessed 20 July 2010

Biggs WD (1960) The brittle fracture of steel. Macdonald and Evans Ltd., London

Charpy G (1904) On testing metals by the bending of notched bars. Int J Fract 1984 25(4):287–305 (Trans from Mémoirs de la Societé de Ingénieurs Civil de France by Towers OL, McSweeney S)

Condurache D, Martinual V (2008) Foucault pendulum-like problems: a tensorial approach. Int J NonLinear Mech 438:743–760

Frost NE, Marsh KJ, Pook LP (1974) Metal fatigue. Clarendon, Oxford

Gere JM, Timoshenko SP (1991) Mechanics of materials. 3rd SI edn. Chapman and Hall, London

Hutzler J, Delaney G, Weaire D, MacLeod F (2004) Rocking Newton's cradle. Am J Phys 72(12):1508–1516

Lamb H (1923) Dynamics, 2nd edn. Cambridge University Press, Cambridge

Matthews RJ (2000) Time for science education. How teaching the history and philosophy of pendulum motion can contribute to science literacy. Kluwer Academic/Plenum Publishers, New York

Matthys R (2004) Accurate clock pendulums. Oxford University Press, Oxford

Phillips N (2005) What makes the Foucault pendulum move among the stars? In: Matthews MR, Gauld CF, Stinner A (eds) The pendulum. Scientific, historical and educational perspectives. Springer, Dordrecht, pp 38–44

Pippard AB (1988) The parametrically maintained Foucault pendulum and its perturbations. Proc R Soc Lond A 420:81–91

Rawlings AL (1993) The science of clocks and watches, 3rd edn. British Horological Institute Ltd., Upton

Schultz-DuBois EO (1970) Foucault pendulum experiment by Kamerlingh Onnes and degenerate perturbation theory. Am J Phys 38(2):173–188

Siewert TA, Manahan MP, McCowan CN, Holt JH, Marsh FJ, Ruth EA (2000) The history and importance of impact testing. In: Siewert TA, Manahan MP (eds) Pendulum impact testing: a century of progress. STP 1380. American Society for Testing and Materials, West Conshohocken, pp 3–16

Synge JL, Griffith BA (1959) Principles of mechanics, 3rd edn. McGraw Hill Book Company Inc., New York

Tipper CF (1962) The brittle fracture story. Cambridge University Press, Cambridge

Tobin W (2003) The life and science of Lon Foucault. The man who proved that the earth rotates. Cambridge University Press, Cambridge

Tobin W, Pippard B (1994) Foucault, his pendulum and the rotation of the earth. Interdiscip Sci Rev 19(4):326–337

Torge W (1980) Geodesy, an introduction. De Gruyter, Berlin (Trans by Jekeli C)

Tóth L, Rossmanith H-P, Siewert TA (2002) Historical background and development of the Charpy test. In: François D, Pineau A (eds) From Charpy to present impact testing. Elsevier, Amsterdam, pp 319 (Charpy Centenary Conference (2001: Poitiers, France))

Tredgold T (1822) A practical essay on the strength of cast iron, intended for the assistance of engineers, iron masters, architects, millwrights, founders, smiths, and others engaged in the construction of machines, buildings, etc, containing practical rules, tables and examples; also an account of some new experiments, with an extensive table of the properties of materials. Taylor J, London

Zevin AA, Filonenko LA (2007) A qualitative investigation of the oscillations of a pendulum with a periodically varying length and a mathematical analysis of a swing. J Appl Math Mech 71(6):892–904

Chapter 7
Engineering

7.1 Introduction

Pendulums are an essential component of some engineering structures. Three of these are described in this chapter. These are the Watt steam governor, cable cars, and tension leg platforms.

The *Watt steam governor*, also known as a *centrifugal governor* or *fly ball governor*, and in French régulateur à boules (ball governor) was invented by James Watt to regulate the supply of steam to his steam engines and hence keep the speed reasonably constant, irrespective of the load (Lineham 1914). A Watt steam governor fitted to a Boulton and Watt steam engine ca. 1810 is shown in Fig. 7.1. It is not clear when James Watt started to use his steam governor. Denny (2002) states that it was in 1788, but the Science Museum in London have a Boulton and Watt steam engine, dated 1781, which is fitted with a Watt steam governor. Other applications for what is then known as the *Watt governor* include windmills, lighthouses and telescope drives (Denny 2002; Tobin 2003). Figure 7.2 shows a Watt governor fitted to a corn mill.

Cable cars are widely used in mountainous areas for transporting people and freight up and down mountains. A cable car has an inline set of grooved wheels that run on a *fixed cable* between two stations (Fig. 7.3). A fixed cable is usually supported by pylons between the stations so that it follows the contour of the mountain (Fig. 7.4). A *gondola* is suspended below the wheel set by a girder that is fixed to the gondola, but pivoted to the wheel set. The gondola is therefore the bob of a pendulum. The pendulum must be correctly designed for satisfactory operation of the cable car.

An important design requirement for the *platforms* used offshore for oil and gas exploration and production is to avoid excitation of structural resonances at *wave passing frequencies*. These are typically in the range 0.08–0.4 Hz (Pook and Dover 1989; Patel and Witz 1991). A *fixed platform* is a structure standing on the sea bed. Avoidance of structural resonances in fixed platforms becomes increasingly difficult as the water depth increases, and they are not normally used if the depth is

L.P. Pook, *Understanding Pendulums: A Brief Introduction*, History of Mechanism and Machine Science 12, DOI 10.1007/978-94-007-1415-1_7,

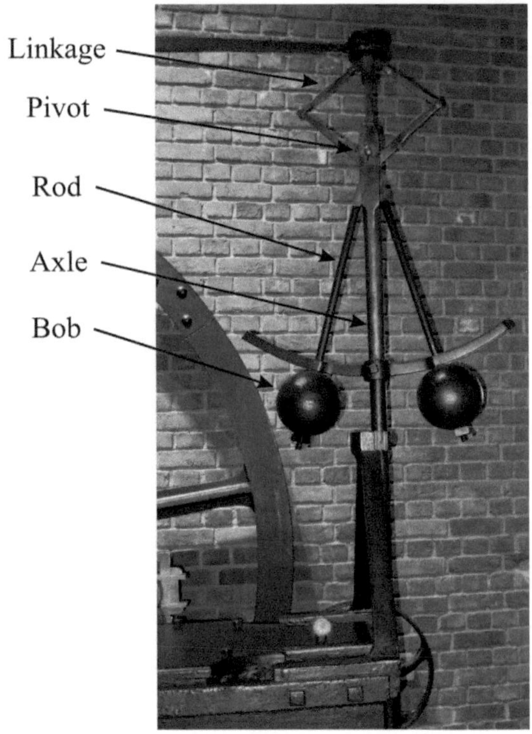

Fig. 7.1 Watt steam governor fitted to a Boulton and Watt steam engine ca. 1810

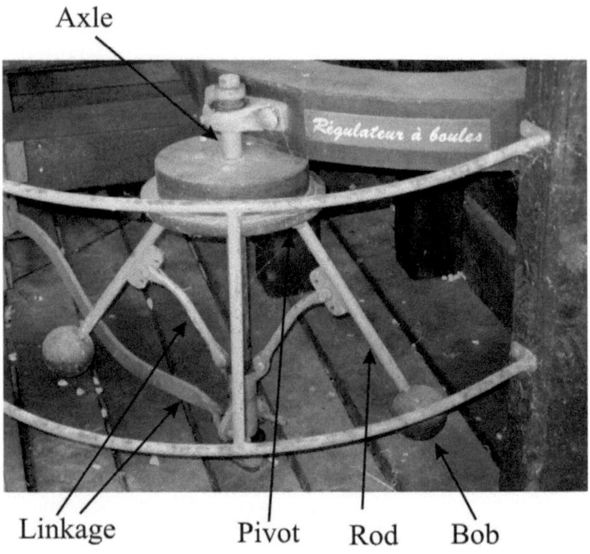

Fig. 7.2 Watt governor fitted to a corn mill

Fig. 7.3 Cable car and station at Aiguille du Midi, France

Fig. 7.4 Cables and pylon for cable cars at Stresa, Italy

greater than about 200 m. For deeper water *tension leg platforms* are now widely used. A tension leg platform is a *floating platform* which is held to the seabed by *tethers*. Tension in the tethers holds the platform below its equilibrium floating level. A tension leg platform is therefore an *inverted pendulum*. A typical tension leg platform is shown schematically in Fig. 7.5. A tension leg platform was first used to produce oil in 1984 (Patel and Witz 1991).

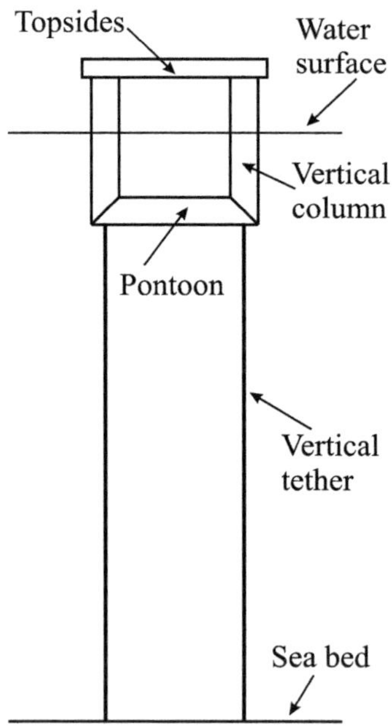

Fig. 7.5 Essential features of a typical tension leg platform

7.2 Watt Steam Governor

Watt steam governors all have the same basic design. There are two identical pendulums, with spherical bobs and light rods attached by a horizontal pivot to a rotating axle driven by the steam engine (Fig. 7.1). The pendulums are real versions of the rotating simple rod pendulum (Fig. 3.7). One of the functions of the associated *linkage* is to ensure that the pendulum angle, θ, shown in Fig. 3.7, is the same for both pendulums. Thus, the governor remains balanced and vibration is avoided. With the governor stationary the pendulum angle is usually about 15° (Fig. 7.1).

The main function of a Watt steam governor, and associated linkage, is to keep the steam engine speed reasonably constant, irrespective of the load (Lineham 1914; Denny 2002). The time of swing, T, for one revolution of the governor is given by Eq. 2.37 where l is the effective length of the pendulums, θ is the pendulum angle and g is the acceleration due to gravity. From Eq. 2.37 (Lineham 1914), the vertical distance of the centre of oscillation of each pendulum below the pivot, l_V (Fig. 3.7) is given by Eq. 3.9 where ω is the angular velocity and is independent of l. Thus, l_V is inversely proportional to ω^2. The linkage transfers the motion of the pendulums to a valve which reduces the supply of steam as the angular velocity increases. Details of the linkage vary widely.

In operation the spherical bobs move out to a pendulum angle, θ, (Fig. 3.7) which depends on the angular velocity of the rotating axle. If the engine speed increases so does the pendulum angle, and the associated linkage reduces the supply of steam to the engine and the engine speed decreases. Conversely, if the engine speed decreases the supply of steam is increased. In practice, the engine speed is not independent of the load; there is always some decrease in engine speed as the load increases. Hence, the governor needs to be made as sensitive as possible to changes in engine speed. However, if the governor is too sensitive the engine speed oscillates about the desired value, this is known as *hunting*. James Clerk Maxwell's theoretical analysis, published in 1868, showed why hunting occurred, and indicated how it could be avoided. Maxwell's analysis is reproduced by Denny (2002). Maxwell's analysis initiated the discipline now known as *control engineering*.

Before Maxwell's analysis was available Watt steam governor designs had to be optimised by trial and error. In particular, the relationship between the angular velocity of the axle and steam valve opening had to be correct. To achieve this, pendulum pivots were sometimes offset from the rotating axle. The offset can be either positive or negative. A positive offset for an idealised Watt steam governor is shown in Fig. 7.6a. The Watt governor fitted to a corn mill, shown in Fig. 7.2, has a positive offset, but this is not clearly visible in the photograph. A Watt steam governor with negative offset (Fig. 7.6b) is known as *Head's governor* (Lineham 1914).

For a positive offset, X, (Fig. 7.6a) the effective pendulum length, l, is increased by $X/\sin\theta$, where θ is the pendulum angle, and there is a *virtual frictionless pivot* above the frictionless pivot. Equation 2.13 for the time of swing, T, for one revolution of the governor becomes

$$T = \frac{1}{2\pi}\sqrt{\frac{l\cos\theta + X\cot\theta}{g}} \tag{7.1}$$

where g is the acceleration due to gravity. Thus, the effect of the positive offset is to change the relationship between the pendulum angle and time of swing. For a

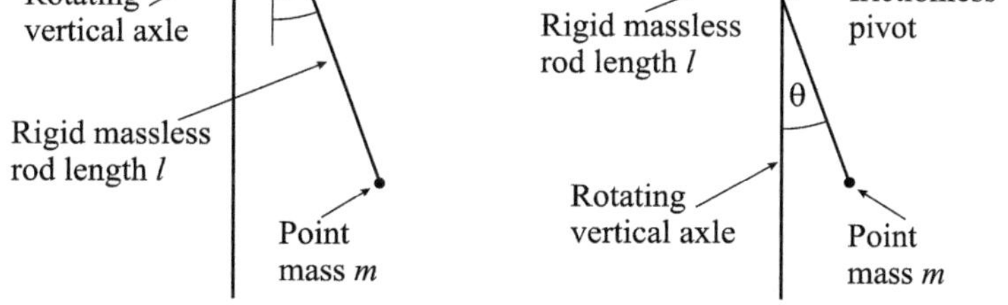

Fig. 7.6 Idealised Watt steam governor pendulums. (**a**) Pivot with positive offset. (**b**) Pivot with negative offset (Head's governor)

given pendulum angle the time of swing is increased, and the relationship between the angular velocity and the vertical distance of the centre of oscillation of each pendulum below the pivot, l_V, is different.

Similarly, for a negative offset (Fig. 7.6b) the virtual frictionless pivot is below the frictionless pivot, and Eq. 7.1 becomes

$$T = \frac{1}{2\pi}\sqrt{\frac{l\cos\theta - X\cot\theta}{g}} \tag{7.2}$$

For a given pendulum angle the time of swing is reduced.

Details of the linkages associated with Watt governors, and Watt steam governors vary widely. Various arrangements were tried in order to optimise the design. In Fig. 7.1 the linkage is above the pivot, and in Fig. 7.2 it is below the pivots.

7.3 Cable Car

Cable cars are widely used in mountainous areas for transporting people and freight up and down mountains. A cable car has an inline set of grooved wheels that run on a *fixed cable* (Fig. 7.7). A *towing cable*, below the fixed cable in the figure, is attached to an *inline wheel set* and moves the cable car along the fixed cable. A fixed cable is usually supported by pylons between the stations so that it follows the contour of the mountain (Fig. 7.4). Cable cars are usually in pairs, attached to the same

Fig. 7.7 Cable car at Aiguille du Midi, France

towing cable so that the cable car going up the mountain is at least partially balanced by the other cable car coming down the mountain. They pass each other midway. The towing cable is driven by an electric motor in one of the stations. A *gondola* (Figs. 7.3 and 7.7) is suspended below the wheel set by a girder that is fixed to the gondola, but pivoted to the bogies carrying the inline wheel set. The bogies are pivoted so that the wheels follow the curvature of the cable, and can pass smoothly over the abrupt changes in cable curvature at pylons (Fig. 7.4). There is always additional curvature due to the weight of the gondola. This additional curvature is visible in Fig. 7.7. If the gondola were suspended from a single wheel it could oscillate in *front-to-back motion* as viewed in the figure, but the cable curvature due to the inline wheel set means that this does not happen.

The girder connecting the inline wheel set to the gondola is shaped so that it clears the fixed and towing cables, and can pass the fixed cable supports at pylons. The girder is pivoted at the wheel set so that it can swing in *left-right motion*. This means that the gondola remains approximately horizontal irrespective of the inclination of the wheel set, and also irrespective of the distribution of passengers and freight within the gondola. It also means that the gondola and girder form a real version of a simple rod pendulum (Sect. 2.3), and can oscillate in left-right motion. Oscillation in left-right motion with a time of swing of a few seconds is clearly noticeable when travelling in a cable car. One of the reasons that the girder is fairly long is that a long time of swing is more comfortable for passengers than a short time of swing. Another reason is that a long girder means that the pendulum angle, θ, (Fig. 2.2) does not alter much when the distribution of passengers and freight within the gondola changes. This is important for the comfort of passengers, and when passengers are embarking and disembarking.

7.4 Tension Leg Platform

Designs of *tension leg platforms* used for oil and gas exploration and production in waters deeper than about 200 m vary widely (Patel and Witz 1991). The essential features of a typical tension leg platform are shown in Fig. 7.5. There are four cylindrical *pontoons*, arranged in a square, and connected by four surface piercing, vertical, cylindrical *columns* to the *topsides*. The accommodation block, production facilities, etc. are mounted on the topsides. The small water line area of the vertical columns minimises the effect of waves on the platform. There is a vertical *tension leg*, also called a *tether*, at each corner connecting the platform to the sea bed. Each tether is a group of wire ropes. The buoyancy (cf Sect. 4.3) of the pontoons and the immersed part of the vertical columns is greater than that needed to support the weight of the platform. Hence, the platform is held in position by tension in the vertical tethers, and it is an *inverted pendulum*. Specifically, the tension leg platform shown in Fig. 7.5 is an inverted real version of a quadrifilar pendulum (Sect. 3.7).

A tension leg platform has two principal modes of oscillation. These are analogous to modes of oscillation of a quadrifilar pendulum. Amplitudes are small, and motions approximate to damped simple harmonic motion (Sect. 4.2.1). The first is a *sway mode of oscillation*, in which the platform remains horizontal, and individual points move on the surface of a sphere. This motion is identical to the motion of the point mass, m, of a simple string pendulum (Sect. 2.4). The second is a torsional mode of oscillation in which the platform remains horizontal and rotates about a vertical axis.

For a particular tension leg platform an effective value of the acceleration due to gravity, g_e, can be defined as

$$g_e = \frac{B}{M} - g \tag{7.3}$$

where B is the buoyancy and M is the mass of the tension leg platform, and g is the acceleration due to gravity. Replacing g by g_e Eq. 2.13 for the time of swing, T, becomes, for the sway mode of oscillation

$$T = 2\pi\sqrt{\frac{l}{g_e}} \tag{7.4}$$

where l is the length of the tethers. Similarly, for the torsional mode of oscillation Eq. 3.12 becomes

$$T = 2\pi\sqrt{\frac{\kappa^2 l}{g_e c^2}} \tag{7.5}$$

where κ is the radius of gyration and $2c$ is the diagonal distance between the tethers.

Times of swing, T, for both modes of oscillation are typically 40 s for 120 long m tethers and 80 s for 500 m tethers (Patel and Witz 1991). Corresponding frequencies are 0.025 and 0.0125 Hz. A tension leg platform is driven by surface waves, in which the water surface elevation, and resulting forces on the vertical columns, approximate to narrow band random processes (Pook and Dover 1989). However, wave passing frequencies are typically in the range 0.08–0.4 Hz (Sect. 7.1). Hence, the spectral density function and resonance curves (Figs. 5.5 and 5.6) do not overlap and, as desired, excitation of oscillations does not occur (Sect. 5.4.2).

References

Denny M (2002) Watt steam governor stability. Eur J Phys **23**(4):339-351

Lineham WJ (1914) A textbook of mechanical engineering, 11th edn. Chapman & Hall, London

Patel MH, Witz JA (1991) Compliant offshore structures. Butterworth-Heinemann, Oxford

Pook LP, Dover WD (1989) Progress in the development of a wave action standard history (WASH) for fatigue testing relevant to tubular structures in the North Sea. In: Watanabe RT, Potter JM (eds) Development of fatigue loading spectra. ASTM STP 1006. American Society for Testing and Materials, Philadelphia, pp 99-120

Tobin W (2003) The life and science of Lon Foucault. The man who proved that the earth rotates. Cambridge University Press, Cambridge

Chapter 8
Entertainment

8.1 Introduction

Pendulums are an essential part of some structures used for entertainment. Three of these uses are described in this chapter. These are child's swings, a child's rocking horse, and pendulum harmonographs. The double rod pendulum (Sect. 3.3), and Newton's cradle (Sect. 6.3) are further examples.

Child's swings are ubiquitous in both public playgrounds and private gardens. One of the attractions is that pumping is possible so they can be ridden without external assistance. Details vary widely, but the main features are shown schematically in Fig. 8.1. A homemade swing in a private garden is shown in Fig. 8.2. This is suspended by ropes from a horizontal branch of a tree.

One form of a *child's rocking horse* is shown in Fig. 8.3. The horse (Fig. 8.3a) is mounted on two wooden bars. These are suspended by pivoted metal links from a fixed wooden bar, which is at the top of the supporting frame. The attraction of the rocking horse is that it can be pumped by the rider to give an exhilarating ride (Fig. 8.3b). The *rocking horse* is, in essence, a four bar linkage with a coupler (Fig. 3.10). The fixed wooden bar corresponds to the static bar, the pivoted metal links at one end to the pilot crank, the two wooden bars on which the horse is mounted to the connecting rod, the pivoted metal links at the other end, to the rocker, and the horse to the coupler. This is shown schematically in Fig. 8.4, which includes key dimensions.

Lissajous figures are plane curves produced when two orthogonal simple harmonic motions (Sect. 2.2) are combined. This can be done either theoretically or experimentally. They are named after Jules Antoine Lissajous, who demonstrated them experimentally in 1857 (Benham 1909; Loney 1913; Lamb 1923; Stong 1965; von Seggern 1990; Ashton 2003). Lissajous used an optical method involving mirrors mounted on two tuning forks arranged at right angles. Kœnig (1865) is a catalogue that includes a commercially available version of Lissajous' apparatus. Some authors describe Lissajous figures without saying that this is what they are, for example Bazley (1875). They were first described by Nathaniel Bowditch in

L.P. Pook, *Understanding Pendulums: A Brief Introduction*, History of Mechanism and Machine Science 12, DOI 10.1007/978-94-007-1415-1_8,

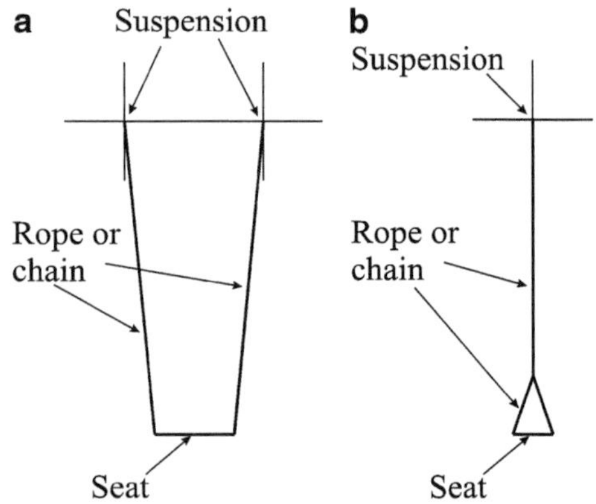

Fig. 8.1 Schematic child's swing. (**a**) Front view. (**b**) Side view

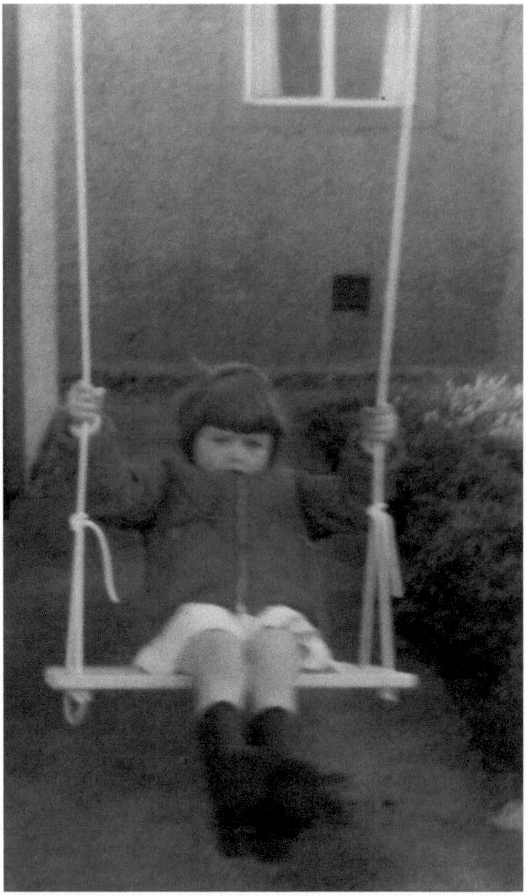

Fig. 8.2 Home made child's swing in a private garden

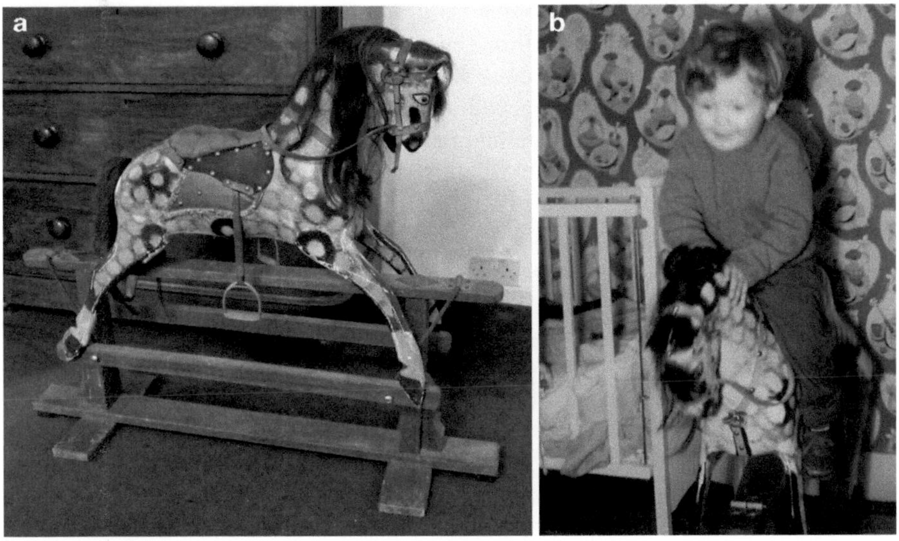

Fig. 8.3 Child's rocking horse. (**a**) General view. (**b**) Being ridden

Fig. 8.4 Schematic child's rocking horse as a four bar linkage with a coupler. In rest position

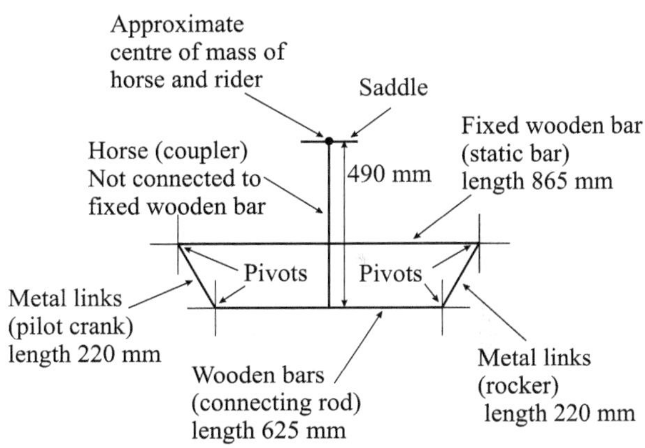

1815 (Greenslade 2010), and they are sometimes called *Bowditch curves* (von Seggern 1990). Lissajous figures are sometimes called *rectangular curves* or *rectilinear harmony curves* because they are produced by compounding two rectilinear motions. Lissajous figures are aesthetically satisfying, and they are closely related to the mathematics of music (Whitty 1893; Newton 1909b; Ashton 2003). Lissajous figures and some related curves are described in this chapter.

A mechanical device for the production of Lissajous figures and some related curves is called a *harmonograph*. This term was in use by 1877 (Whitaker 2005). There are two main types of harmonograph. A *circular harmonograph* is based on the rotation of interconnected wheels or gears, and numerous versions have been described (Kœnig 1865; Anonymous 1930a; Stong 1965; van den Berg 1997; Whitaker 2001b; Anonymous 2010b). The geometric chuck and lathe described by Bazley (1875) is, in effect, a circular harmonograph. *Pendulum harmonographs* are based on the oscillations of pendulums. Curves are close approximations to theoretical ideals provided that pendulum amplitudes are small. Because of friction etc. (Sect. 4.2) the amplitude of oscillation decays with time. Hence, the curves produced become progressively smaller.

There are many different types of *pendulum harmonograph* (Newton 1909b; Anonymous 1930a, b, 1955; Stong 1965; Cundy and Rollett 1981; Whitaker 2001a, 2005; Ashton 2003). Descriptions can also be found on the Internet, although these are of varying quality. The earliest pendulum harmonographs were based on the Blackburn pendulum (Sect. 3.4), which was first discussed by James Dean in 1815 (Whitaker 1991, 2005). It appears to have been re-discovered independently by Blackburn, who first described it in 1844 (Benham 1909; Ashton 2003). *Twin pendulum harmonographs*, which use two pendulums suspended by gimbals so that they are free to swing in any direction, appeared in the 1870s (Whitaker 2001a). The first of these was *Tilley's compound pendulum*, ca 1873 (for 'compound' read 'twin'). Commercial versions of twin pendulum harmonographs started to become available at about this time. In 1906 Charles Benham introduced his *twin-elliptic harmonograph* (Whitaker 2001a). A schematic twin-elliptic harmonograph is shown in Fig. 8.5a, and a Meccano version in Fig. 3.14. The Meccano version is based on a design by Anonymous (1955). Both parts of the pendulum of a twin elliptic harmonograph are free to swing in all directions. The upper part of the double pendulum is suspended by a gimbal, and is sometimes called the *main pendulum*, and the lower part, suspended by a string from the main pendulum, is sometimes called the *deflector pendulum*. Some other types of pendulum harmonograph are reviewed by Whitaker (2001a).

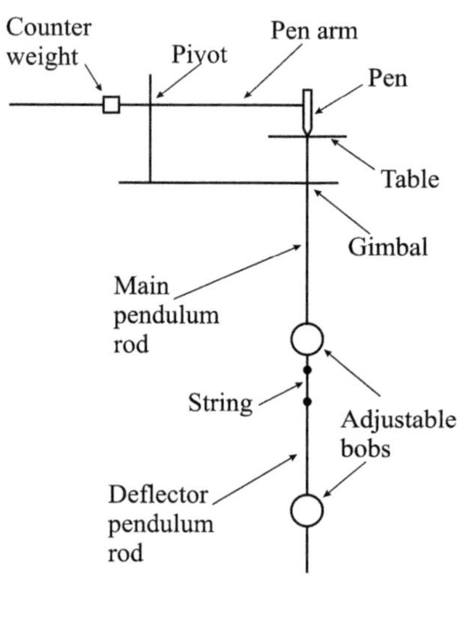

Fig. 8.5 Schematic twin-elliptic harmonograph

A harmonograph always has a *recording device*. This is usually a sheet of paper fastened to a moving table (Fig. 8.5). A pen is mounted on a pivoted pen arm so that it remains in contact with the paper. The pen arm has an adjustable counter weight so that the pressure between pen and paper can be adjusted. The pen can be held clear of the table while the paper is being changed and the harmonograph started.

The figures produced by harmonographs are sometimes called *harmonograms*. The aesthetic quality of harmonograms produced by pendulum harmonographs became appreciated towards the end of the nineteenth Century, and their production was a popular pastime (Cundy and Rollett 1981; Ashton 2003). At about the same time the relationship between Lissajous figures and the theory of music became of interest, and is discussed by Whitty (1893), Benham (1909), Goold (1909) and Ashton (2003).

8.2 Child's Swings

The main features of a *child's swing* are shown in Fig. 8.1. A flat rectangular seat is supported at each corner by either ropes or chains, and the swing is suspended from a suitable support. Ropes are usually used in private gardens (Fig. 8.2), but chains are usually used in public gardens. The extra weight of the chains makes these swings more difficult to pump. Viewed from the front (Fig. 8.1a) the ropes or chains slope outwards towards the support. Viewed from the side the ropes or chains are arranged in an inverted Y-shape (Fig. 8.1b).

A child's swing approximates to a real version of a dual string pendulum (Sect. 3.7.1), It has three degrees of freedom and three modes of oscillation. In the principal mode of oscillation, which is the desired mode of oscillation, the centre of oscillation of the swing and rider moves in an arc in a vertical plane in *front-to-back motion*, as viewed from in front of the rider. Amplitudes are large so the motion does not approximate to simple harmonic motion. In the first secondary mode of oscillation the swing and rider swing in *left-right motion*. The second secondary mode of oscillation is a torsional mode of oscillation in which the seat and rider rotate about a vertical axis.

For an entertaining ride, a child's swing should swing as freely as possible in the principal mode of oscillation, and the secondary modes of oscillation should be suppressed as much as possible. The secondary modes of oscillation can be suppressed by inclining the ropes or chains outward. The left-right motion is then the same as the motion of a trapezium pendulum (Sect. 3.8). For the home made child's swing shown in Fig. 8.2 the inclination is critical for stable oscillation in the principal mode of oscillation, and was found by trial and error to be about 4°. The arrangements of the ropes or chains in an inverted Y-shape (Fig. 8.1b) ensures that the seat remains at a comfortable angle.

8.2.1 Pumping of Child's Swings

Child's swings are not powered, so *pumping* (Sect. 6.4.2) is needed to set them in motion and keep them swinging. In *rider pumping* the rider changes position in time with the motion of the swing. Rider pumping has been discussed by several authors, including Tea and Falk (1968), Case and Swanson (1990), Case (1996), Roura and González (2010), and Yang et al. (2010). In *external pumping* the rider takes no action, and someone else pumps the swing. In general, mathematical analyses of pumping only lead to qualitative conclusions because of the difficulty of selecting appropriate input data. External pumping of child's swings is by someone standing behind the rider who pushes the swing at the rearward end of each front-to-back swing. External pumping results in *angular pumping* (Sect. 6.4.2), with one positive impulse per cycle, as shown in the phase diagram, Fig. 8.6a.

Rider pumping of child's swings can be divided into different types, although in practice a combination of more than one type of pumping is normally used. In *rocking rider pumping*, which can be either in the standing position (Case 1996), or seated position (Case and Swanson 1990; Roura and González 2010), the rider pulls and pushes on the ropes or chains to rock backwards at the rearward end of each front-to-back swing and forward at the forward end of each front-to-back swing. Starting a child's swing from the rest position by rocking rider pumping is difficult. Rocking rider pumping results in angular pumping, with two positive impulses per cycle, as shown in the phase diagram, Fig. 8.6b. In *leg swinging rider pumping* in the seated position, the legs are swung forward at the rearward end of each front-to-back swing, and backwards at the forward end of each front-to-back swing. Starting a child's swing from the rest position by leg swinging pumping is possible but difficult. Leg swinging rider pumping results in angular pumping, with two positive impulses per cycle, as shown in the phase diagram, Fig. 8.6b.

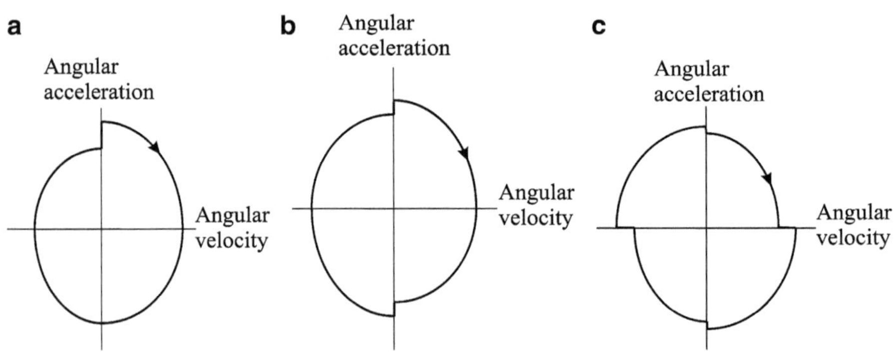

Fig. 8.6 Phase diagrams for pumping of swings. *Arrows* show the sense of increasing time. (**a**) External pumping of a swing. (**b**) Rocking rider pumping and leg swinging rider pumping, of a child's swing, and rocking rider pumping of a child's rocking horse. (**c**) Squatting pumping of a child's swing

In *squatting rider pumping* in the standing position (Tea and Falk 1968; Yang et al. 2010), the rider squats downwards at each end of a front-to-back swing, and stands upright as the child's swing passes through the rest position. The changes of position of the rider decrease the effective length of the swing (pendulum) as it passes through the rest position, and increase the effective length at the ends of the front-to-back swings. Starting a child's swing from the rest position by squatting pumping is impossible. Squatting rider pumping results in a form of radial pumping, with two positive and two negative impulses per cycle, as shown in the phase diagram, Fig. 8.6c. The effect of centripetal force (Sect. 6.3.2) as the swing passes through the rest position means that more energy is provided by the positive impulses than is extracted by the negative impulses, as shown schematically in the figure. Squatting rider pumping results in *vertical pumping*. It differs from pumping a pendulum by moving the suspension point up and down (Sect. 6.4.2), but the effects are similar.

8.3 Child's Rocking Horse

In essence the *child's rocking horse* shown in Fig. 8.3 is a four bar linkage with a coupler (Fig. 3.10), so has one degree of freedom. The relationship is shown schematically in Fig. 8.4. The only mode of oscillation is in front-to back motion, as viewed from in front of the rider. The centre of mass of the horse and rider is above the pivots so the child's rocking horse is an inverted version of a *trapezium pendulum* (Fig. 3.11).

The centre of mass of the horse and rider is lowest in the rest position. Away from the rest position the centre of mass moves in a non circular arc and there is a *virtual pivot*, which is not in a fixed position, above the centre of mass. There is no simple expression for the path traced by the centre of mass (Sect. 3.8). A rocking horse is pumped by rocking rider pumping (previous section). The rider rocks backwards at the rearward end of each front-to-back swing and forwards at the forward end of each front-to-back swing (Fig. 8.3b). Rocking rider pumping results in angular pumping, with two positive impulses per cycle, as shown in the phase diagram, Fig. 8.6b. The dimensions (Fig. 8.4) are chosen so that even with vigorous pumping it is difficult to make the horse swing beyond positions in which either the wooden bars (connecting rod) and metal links (pilot crank) are in line (Fig. 8.7), or the wooden bars (connecting rod) and metal links (rocker) are in line. This effectively limits the amplitude of swings and prevents overturning.

8.4 Pendulum Harmonographs

There are many different types of *pendulum harmonograph* (Sect. 8.1). All are designed so that the amplitude of oscillation of the pendulums is small. A pendulum harmonograph usually has two pendulums that are individually adjustable.

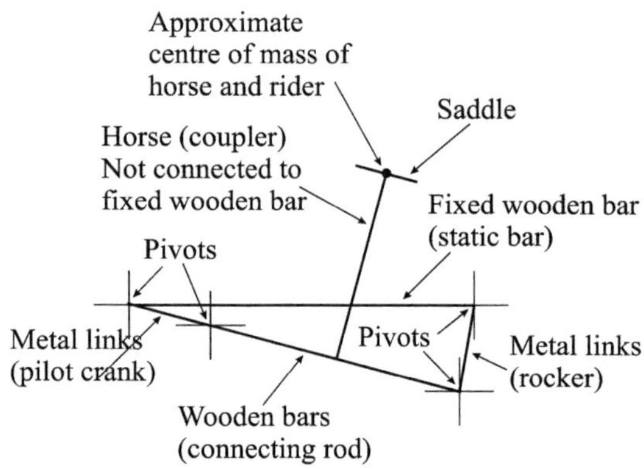

Fig. 8.7 Schematic child's rocking horse as a four bar linkage with a coupler. With metal links (pilot crank) and wooden bars (connecting rod) in line

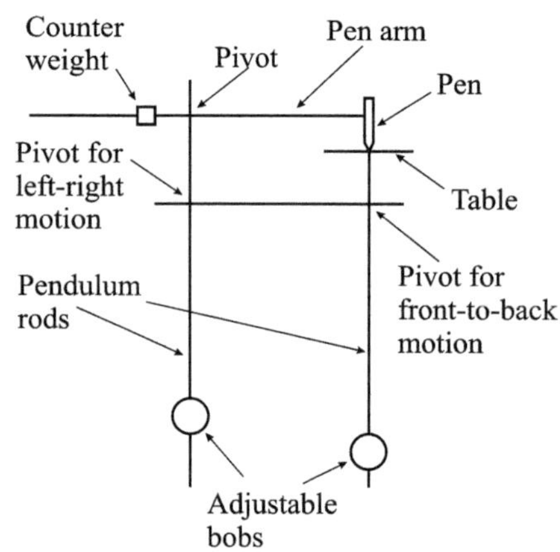

Fig. 8.8 Schematic simple twin pendulum harmonograph

An exception is *Benham's triple pendulum* that has three (Benham 1909). Some pendulum harmonographs include arrangements so that times of swing can be easily and precisely adjusted to facilitate the production of desired types of harmonogram. Arrangements to ensure appropriate launching of the pendulums are also sometimes included. They are sometimes separate, so the time of swing, T, of individual pendulums is given approximately by Eq. 2.13 where l is the effective length of a pendulum and g is the acceleration due to gravity. In the *twin-elliptic harmonograph* (Figs. 3.14 and 8.5) the pendulums are arranged in series as a double pendulum (Sects. 3.3 and 3.9). In general, there are no simple expressions for times of swing of a twin-elliptic harmonograph. In a pendulum harmonograph inevitable losses, due to friction etc., mean that it is a dissipative system and, in the absence of any impulses, the pendulum amplitude gradually decays with time (Sect. 4.2). Examination of published harmonograms suggests that the pendulum quality, Q, is typically in range the 30–70, which is low by horological standards (Sect. 4.2.2).

Pendulum harmonographs can be conveniently classified in terms of their degrees of freedom. They usually have either two or four degrees of freedom. A *simple twin pendulum harmonograph* is shown schematically in Fig. 8.8. Each pendulum

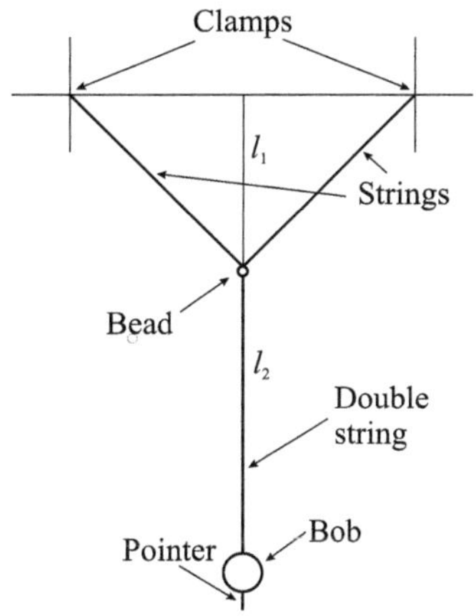

Fig. 8.9 Schematic Blackburn harmonograph

is suspended by a pivot so that it can only swing in one direction. These directions are usually at right angles, but other angles may be used. Each pendulum is a real version of a simple rod pendulum (Sect. 2.3), each of which has one degree of freedom, so a simple twin pendulum harmonograph has two degrees of freedom. The *Blackburn harmonograph*, shown schematically in Fig. 8.9, is based on the Blackburn pendulum (Sect. 3.4), and has two degrees of freedom. The bead can be moved up and down the strings to change the relative values of l_1 and l_2 and hence the relative values of the times of swing of the two modes of oscillation, as given by Eqs. 3.8 and 3.9. It is difficult to arrange for a Blackburn harmonograph to draw a line on paper, and so record the motion of the bob. One possibility for an alternative recording device is to spread sand below the bob so that it is progressively marked by the pointer as it passes (cf Sect. 6.4.1).

In a *twin pendulum harmonograph* both pendulums are suspended by gimbals so they are spherical rod pendulums (Sect. 2.5) with two degrees of freedom. Hence, a twin pendulum harmonograph has four degrees of freedom. Various arrangements are possible. A *Newton's twin pendulum harmonograph* (Newton 1909a) is shown schematically in Fig. 8.10. The simple twin pendulum harmonograph shown schematically in Fig. 8.8 is a Newton's twin pendulum harmonograph with the gimbals replaced by orthogonal pivots. The *twin-elliptic harmonograph* (Figs. 3.14 and 8.5) is equivalent to the double string pendulum (Sect. 3.9), and has four degrees of freedom.

8.5 Harmonograms and Pendulum Harmonographs

The raison d'être of a *pendulum harmonograph* is the production of *harmonograms*. The curves that make up harmonograms can all be regarded as variants of Lissajous figures, described below. There are two approaches to using a pendulum

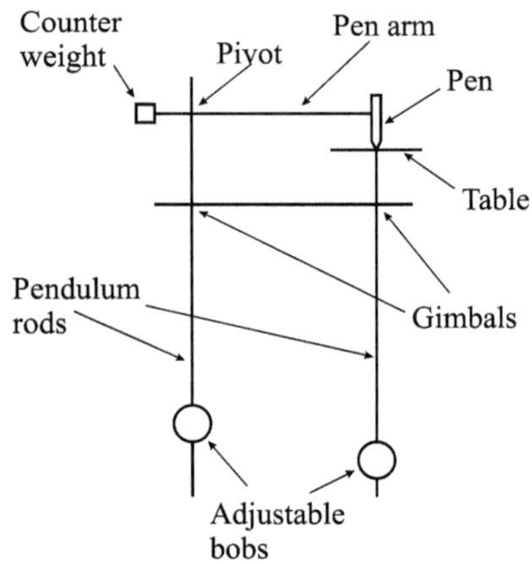

Fig. 8.10 Schematic Newton's twin pendulum harmonograph

harmonograph to produce harmonograms. One approach is to set up and launch the pendulum harmonograph in a random manner, and see what happens. Anonymous (1930b) includes some random harmonograms obtained using a twin-elliptic harmonograph. A more sophisticated approach is to try and create desired types of harmonograms. Whitty (1893) and Ashton (2003) include extensive, organised, collections of harmonograms obtained using pendulum harmonographs. Both authors emphasise the relationship between the mathematics of the harmonograms presented and the mathematics of music. Bazley (1875) includes 3,500 harmonograms obtained by using a geometric chuck as a circular harmonograph. He gives the gear train used to produce each harmonogram, but their mathematics is not discussed.

8.5.1 Lissajous Figures

Lissajous figures are plane curves produced when two orthogonal simple harmonic motions (Sect. 2.2) are combined. For the mathematical construction of Lissajous figures, versions of Eq. 2.2 for the two simple harmonic motions can be conveniently written in the parametric form (von Seggern 1990)

$$x = \sin(at + b\pi) \tag{8.1}$$

$$y = \sin t \tag{8.2}$$

where a is the ratio of the periods of the two simple harmonic motions, and b is a measure of their phase. For a Lissajous figure to be a closed curve a must be rational. If desired, an additional parameter can be introduced into Eq. 8.1 to cover the case where the two simple harmonic motions have different amplitudes. Other, equivalent, equations are sometimes used.

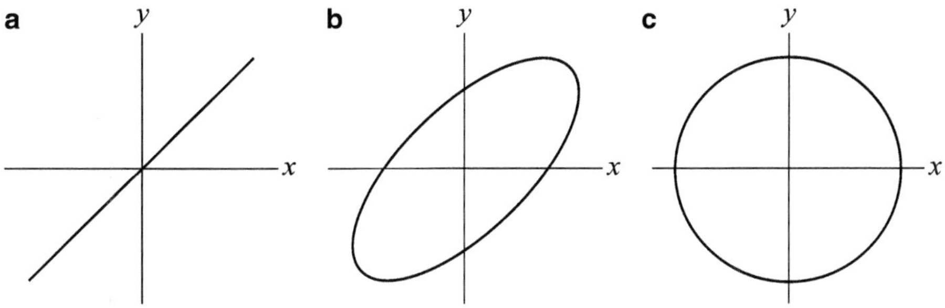

Fig. 8.11 Lissajous figures for equal frequencies and amplitudes. (**a**) Phase 0°. (**b**) Phase 45°. (**c**) Phase 90°

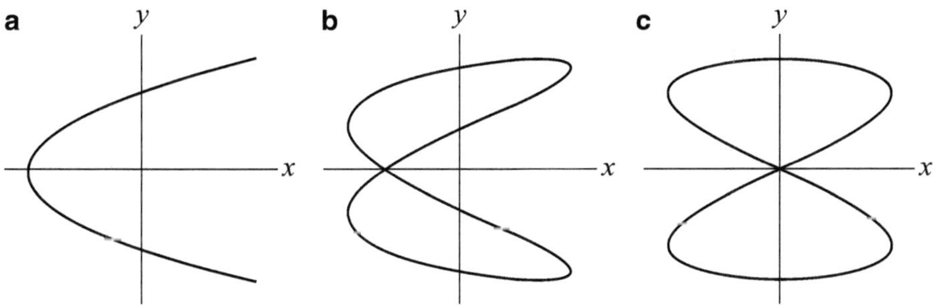

Fig. 8.12 Lissajous figures for horizontal frequency twice the vertical frequency and equal amplitudes. (**a**) Phase 0°. (**b**) Phase 45°. (**c**) Phase 90°

Figures 8.11 and 8.12 show some simple Lissajous figures calculated using Eqs. 8.1 and 8.2. Figure 8.11 is for simple harmonic motions of equal frequency. In general, the Lissajous figure is an ellipse (Fig. 8.11b). For a phase of 0° it appears to be a straight line (Fig. 8.11a). The straight line is actually a *digon*. A digon is a regular polygon which has two vertices and two coincident sides of equal length (Coxeter 1961). For a phase of 90° the Lissajous figure is a circle (Fig. 8.11c), and the angular velocity is constant. Figure 8.12 is for a horizontal (x-axis) frequency of twice the vertical (y-axis) frequency. The ratio between these two frequencies, chosen to be >1, is the *relative frequency*, $1/a$, and is 2. A Lissajous figure can be traversed in either direction so Lissajous figures occur in enantiomorphic (mirror image) pairs. The two parts of an *enantiomorphic pair* are not usually regarded as distinct. If the relative frequency is irrational then the phase is indeterminate, and a Lissajous figure is a *plane filling curve* that, for equal amplitudes, fills a square as $t \to \infty$. The square is the *envelope* of the curve.

For Lissajous figures definitions of *phase angle, in phase*, and *out of phase* differ from those used in Sect. 3.3. Phase angles for equal frequencies are shown in Fig. 8.11. When the two frequencies are different and the relative frequency is rational, then the *beat frequency* is the difference between the two. A given configuration of the two simple harmonic motions then repeats after one *beat cycle*. The beat cycle can be represented by a rotating vector, as in Fig. 2.6. A consistent definition of the phase, ϕ, of a Lissajous figure can be obtained by definition of ϕ for two extreme cases as follows. A Lissajous figure is in phase ($\phi = 0°$) when cusps appear, and this is sometimes called the *cusped phase*, for example Fig. 8.12a. A Lissajous figure is out of phase ($\phi = 90°$) when there are configurations where both simple harmonic motions are at the rest position. This is sometimes called the *open phase* because of the open appearance of a Lissajous figure (Benham 1909), for example Fig. 8.12c. Intermediate phases have an intermediate appearance, for example Fig. 8.12b. The *relative frequency*, f_R, of the component simple harmonic motions can be determined by counting the loops in an open phase. Thus, in Fig. 8.12c there are two loops in a vertical direction and one loop in a horizontal direction, so $f_R = 2/1 = 2$. Three parameters are needed to define the form of a Lissajous figure. These are the relative frequency, the relative amplitude and the phase. Two additional parameters are needed to define its magnitude and orientation.

Benham (1909) gives instructions for the graphical construction of Lissajous figures. They can also be synthesised electronically using an oscilloscope, and be computer generated. There are interactive websites where the user can generate Lissajous figures. For example, see Anonymous (2010b) for a computer generated simulation of an oscilloscope. There are videos on YouTube of mechanically generated, electronically synthesised, and computer generated Lissajous figures (Anonymous 2010a).

8.5.1.1 Damped Lissajous Figures

The combination of two orthogonal damped harmonic motions (Sect. 4.2.1) results in a *damped Lissajous figure* which becomes progressively smaller with time, and therefore has a spiral form. For example, the Lissajous figure for equal frequencies and amplitudes, phase 90° shown in Fig. 8.11c, which is a circle, becomes the corresponding damped Lissajous figure, shown in Fig. 8.13, which is an equiangular spiral that continues inwards indefinitely. The arrow shows the sense of time.

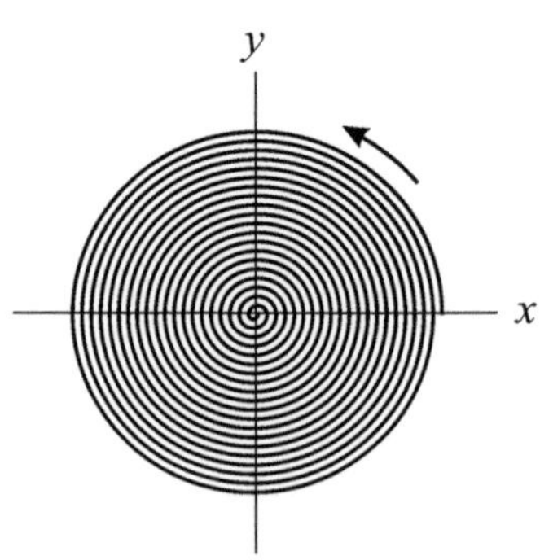

Fig. 8.13 Damped Lissajous figure for equal frequencies and amplitudes, phase 90°. The *arrow* shows the sense of time

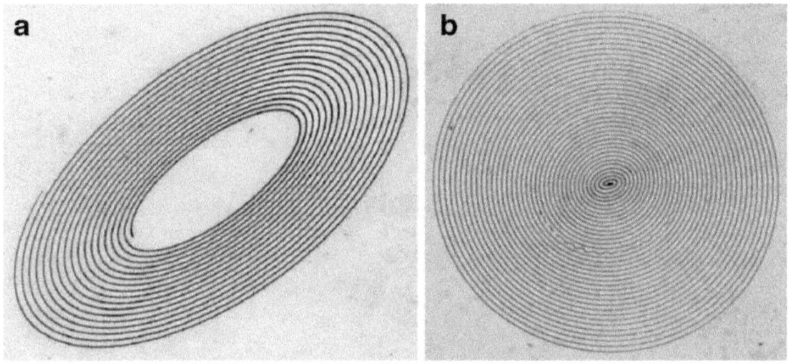

Fig. 8.14 Harmonograms for equal pendulum frequencies. (**a**) Phase 45°. (**b**) Phase 90° (open phase) (Whitty 1893)

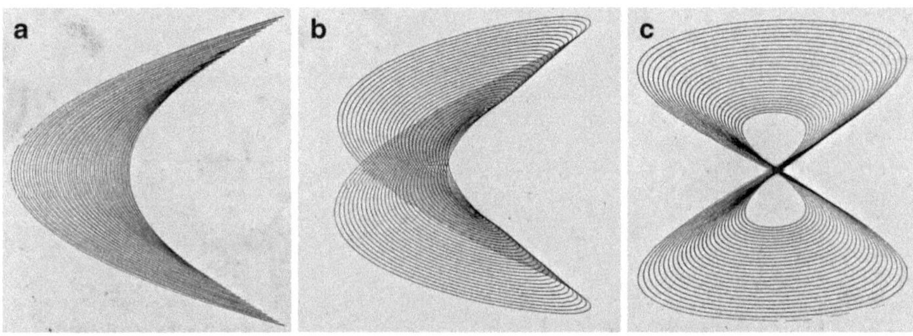

Fig. 8.15 Harmonograms for horizontal pendulum frequency twice the vertical frequency, relative frequency 2. (**a**) Phase 0° (cusped phase). (**b**) Phase 45°. (**c**) Phase 90° (open phase) (Whitty 1893)

The production of harmonograms that approximate to damped Lissajous figures requires a pendulum harmonograph, with two degrees of freedom and motions in orthogonal directions. The simple twin pendulum harmonograph shown in Fig. 8.8 could be used. Pendulum harmonographs with more degrees of freedom can be used provided that the pendulums are correctly launched. The harmonograms shown in Figs. 8.14–8.17 are typical of harmonograms that approximate to damped Lissajous figures. They were obtained by Whitty (1893). He used a twin pendulum harmonograph fitted with an electromagnetic device to ensure correct launching of the pendulums. The pendulum amplitude decay, illustrated by the harmonograms in the figures, is analogous to the fading of musical notes produced by plucked or struck strings in musical instruments (Ashton 2003). A damped Lissajous figure repeats on a smaller scale. It is an example of a *fractal object*, whose appearance is independent of the scale at which it is viewed (Falconer 1990). In practice, pendulums eventually stop swinging so a harmonogram that approximates to a damped Lissajous figure is only a fractal object over a limited range of scales.

In Fig. 8.14 the pendulum frequencies are equal (cf Fig. 8.11), and the amplitudes are approximately equal, so the harmonograms are approximately damped versions

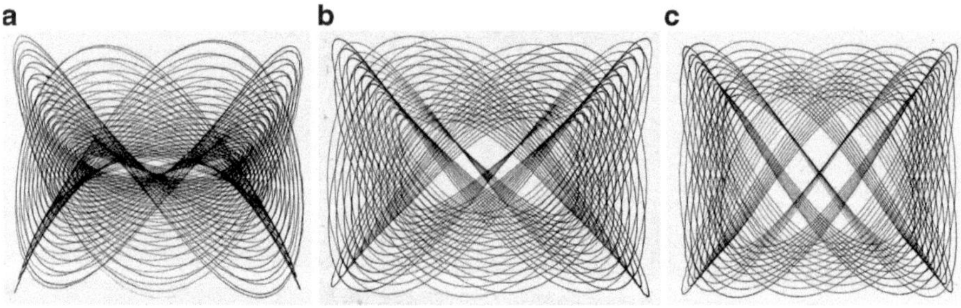

Fig. 8.16 Harmonograms for relative frequency 1.6. (**a**) Phase 0° (cusped phase). (**b**) Phase 45°. (**c**) Phase 90° (open phase) (Whitty 1893)

Fig. 8.17 Harmonograph for relative frequency 1.6, phase 90° (open phase), one beat cycle (Whitty 1893)

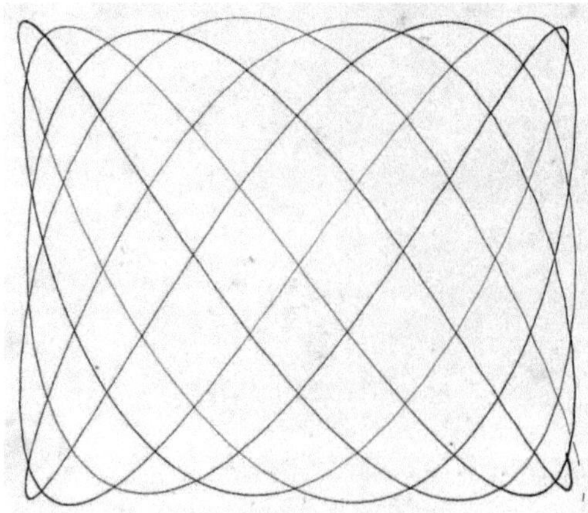

of the Lissajous figures shown in Fig. 8.11b, c. Figure 8.14b is, as would be expected, similar to Fig. 8.13. In musical terminology this is *unison* where two notes have the same frequency (Whitty 1893; Ashton 2003). In Fig. 8.15 the horizontal direction corresponds to a pendulum with twice the frequency of the pendulum which corresponds to the vertical direction, and the amplitudes are approximately equal. The interval between the two frequencies is an *octave*. The ratio between these two frequencies, chosen to be >1, is the *relative frequency*, and is 2. The harmonograms are approximately damped versions of the Lissajous figures shown in Fig. 8.12. Lissajous figures for various musical intervals are tabulated by Ashton (2003).

In general, the simpler the appearance of a Lissajous figure, the more pleasing is the harmony between corresponding musical notes (Whitty 1893; Ashton 2003). In Fig. 8.16 the approximate damped Lissajous figures are complicated. The amplitudes are approximately equal, and the relative frequency is 1.6. In musical terminology the interval is a *minor sixth*, which is a harsh interval. In the open phase (Fig. 8.16c) there are eight loops in a horizontal direction, and five loops in

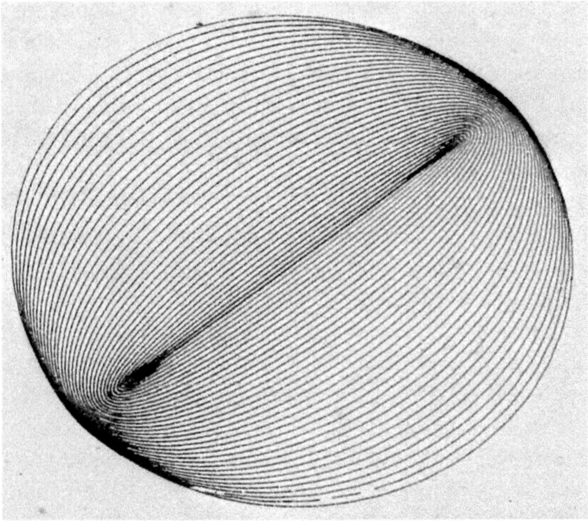

Fig. 8.18 Harmonogram for slightly different pendulum frequencies. Initial phase 90° (Whitty 1893)

a vertical direction. This confirms that the relative frequency is $8/5 = 1.6$. The loops are clearer in Fig. 8.17 where the harmonogram was discontinued after one beat cycle. The *envelope* of the harmonogram is a rectangle. The ratio of the lengths of its sides is the *relative amplitude* of the two pendulums.

A harmonogram for slightly different pendulum frequencies and equal amplitudes is shown in Fig. 8.18. The initial phase was 90°, but this has gradually changed to 0°, that is from an open phase to a cusped phase. If left to continue the pendulums would have gradually changed to an open phase, and so on. In musical terminology, this phenomenon is called *beating*, The frequency with which they move in and out of phase is the *beat frequency*. Done deliberately, beating can add to the richness of musical sound.

8.5.2 Circular Harmony Curves

A *circular harmony curve* is produced by the combination of two circular motions in a plane. The angular velocity of each circular motion is constant. If the rotations are in the same direction (*concurrent rotation*), the result is an *epitrochoid*. If the rotations are in the opposite direction (*countercurrent rotation*) the result is an *hypotrochoid*. Countercurrent rotation is sometimes called *antagonistic rotation*. Equations for epitrochoids and hypotrochoids are complicated and are given by von Seggern (1990) and Whittaker (2001b). Some epitrochoids and hypotrochoids are given special names, but there are inconsistencies so none are given here.

Circular motion with constant angular velocity is a combination of two simple harmonic motions so a circular harmony curve is the result of a combination of four simple harmonic motions. Benham (1909) describes a graphical method for the construction of circular harmony curves. He also discusses the relationship of countercurrent circular harmony curves to regular star polygons. Three parameters are needed to define the form of a circular harmony curve. These are the relative frequency, the relative amplitude and whether rotation is concurrent or countercurrent. Two additional parameters are needed to define its magnitude and orientation. The phase between the two circular motions does not affect the appearance of circular harmony curves, but it does affect their orientation.

Some circular harmony curves are shown in Figs. 8.19–8.21. *Relative frequencies* are shown as ratios, expressed in lowest terms. For concurrent rotation the number of external loops is the difference between the terms of the relative frequency, and for countercurrent rotation it is the sum. Care is needed in interpretation. Circular harmony curves for equal amplitude, relative frequency 2:1, are shown in Fig. 8.19. As expected, the number of external loops is 1 for concurrent rotation (Fig. 8.19a), and is 3 for countercurrent rotation (Fig. 8.19b).

The appearance of a circular harmony curve is strongly dependent on the *relative amplitude*, which is the amplitude of the faster rotation divided by the amplitude of the slower rotation. For example, circular harmony curves for relative amplitude 1:2, relative frequency 2:1, are shown in Fig. 8.20. For concurrent rotation (Fig. 8.20a) the internal loop in Fig. 8.19a has become an inward pointing vertex. For

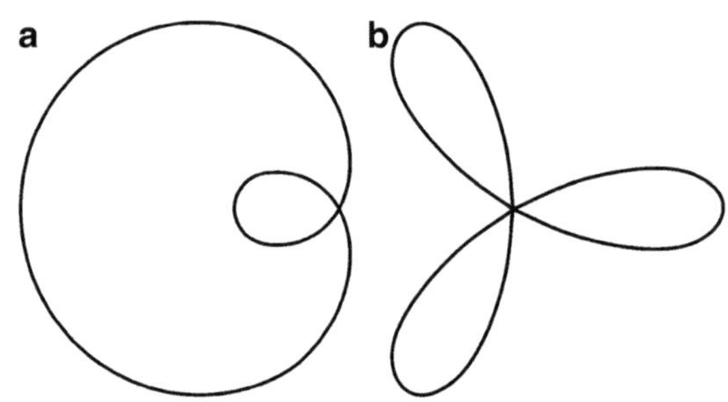

Fig. 8.19 Circular harmony curves for equal amplitude, relative frequency 2:1. (**a**) Concurrent rotation. (**b**) Countercurrent rotation

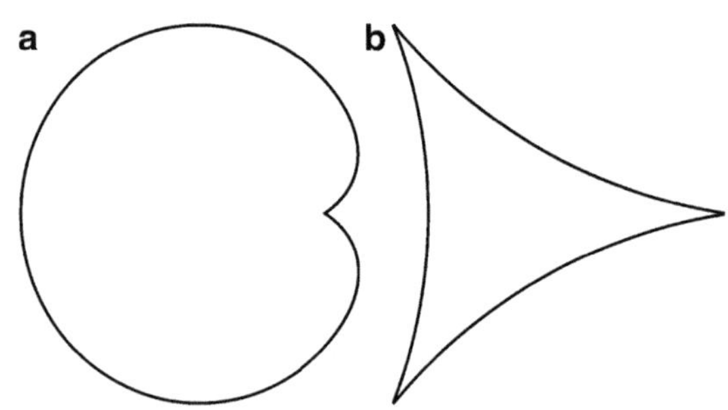

Fig. 8.20 Circular harmony curves for relative amplitude 1:2, relative frequency 2:1. (**a**) Concurrent rotation. (**b**) Countercurrent rotation

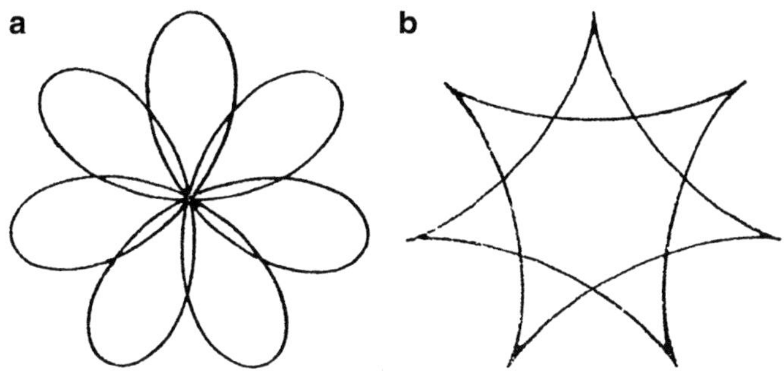

Fig. 8.21 Circular harmony curves for relative frequency 5:2, counter current rotation. (**a**) Equal amplitude. (**b**) Relative amplitude 2:5 (Bazley 1875)

Fig. 8.22 A star heptagon

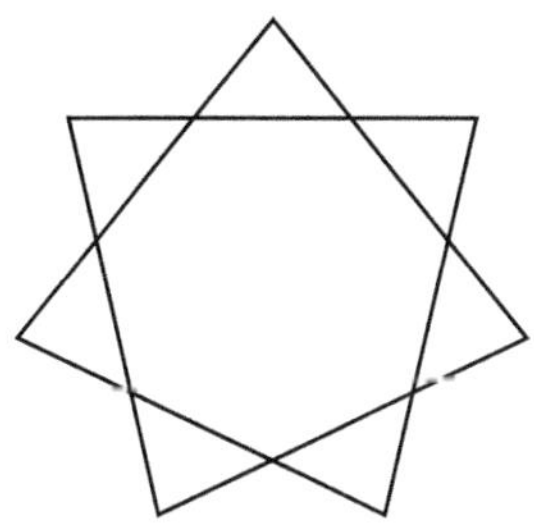

countercurrent rotation (Fig. 8.20b) the external loops in Fig. 8.19b have become outward pointing vertices. Thus, in the relationship with relative frequency, outward pointing vertices are equivalent to external loops.

Goold (1909) gives additional data for some of Bazley's harmonograms that are circular harmony curves. Two of Baxley's harmonographs are shown in Fig. 8.21. Both are for relative frequency 5:2, counter current rotation. For equal amplitudes (Fig. 8.21a) there are seven loops, as expected since $5+2=7$. For relative amplitude 2:5 the loops have become vertices (Fig. 8.21b). The harmonogram is equivalent to the regular star heptagon shown in Fig. 8.22 with the straight edges replaced by non circular arcs. The are there are circular harmony curves equivalent in this way to all regular polygons. That shown in Fig. 8.20b is another example, it is equivalent to an equilateral triangle.

Further examples of circular harmony curves are given by Goold (1909). Benham (1909), Goold (1909) and Ashton (2003) give examples of harmonograms that are approximately damped circular harmony curves. Production of the latter requires a harmonograph with four degrees of freedom.

8.5.3 Miscellaneous Harmonograms

Other combinations of the motions represented by the Lissajous figures shown in Fig. 8.11 can also lead to harmonograms. For example, simple harmonic motion

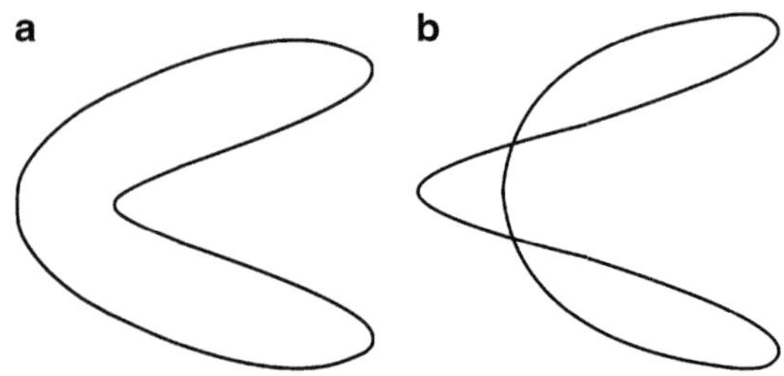

Fig. 8.23 Elliptical harmony curves for combined elliptical motions of equal amplitude and phase, relative frequency 2:1, open phase. (**a**) Concurrent rotation. (**b**) Countercurrent rotation

(Fig. 8.11a) combined with circular motion (Fig. 8.11c) is shown in Fig. 6.9. The frequency of the simple harmonic motion is very high compared with that of the circular motion.

An *elliptical harmony curve* is produced by the combination of two elliptical motions in a plane. An example is given by Benham (1909). Figure 8.23 shows elliptical harmony curves for combined elliptical motions of equal amplitude and phase. These were derived by using the graphical method of Benham (1909). The major axes of the ellipse are orthogonal and their relative frequency is 2:1. Figure 8.23a is for concurrent rotation, and Fig. 8.23b is for countercurrent rotation.

Figures 8.24 and 8.25 show some of the harmonograms obtained by random launching of the Meccano twin-elliptic harmonograph shown in Fig. 8.5. This harmonograph has provision for addition of weights to the upper and lower bobs. For Fig. 8.24 1.5 lb (680 g) was added to the upper bob, and 0.5 lb (227 g) was added to the lower bob. For Fig. 8.25 0.5 lb (227 g) was added to both bobs. Random launching has resulted in a wide variety of harmonograms. The harmonograms in Fig. 8.25 have a more open appearance than those in Fig. 8.24.

8.5.4 Some Practical Considerations

Good quality *pendulum harmonographs* were at one time available commercially, but are not now available. Instructions for their construction are given by several authors, including Anonymous (1955) and Cundy and Rollett (1981). Instructions are also available on the Internet, but these are of varying quality. A pendulum harmonograph is sensitive to extraneous tremors and needs to be firmly mounted. Extraneous oscillations due to a tremor can be seen near the centre of Fig. 8.14b. If the motions are not to decay too rapidly heavy pendulum bobs and low friction between the pen and paper are needed. In other words, pendulum quality (Sect. 4.2.2) needs to be high. A modern, partly used, rollerball pen was used to produce the harmonograms shown in Figs. 8.24 and 8.25. Friction is low, but there was some skipping.

Instructions for the operation of pendulum harmonographs are given by several authors including Benham (1909), and Cundy and Rollett (1981). If a particular harmonogram is desired, then the pendulums must be adjusted to give the correct

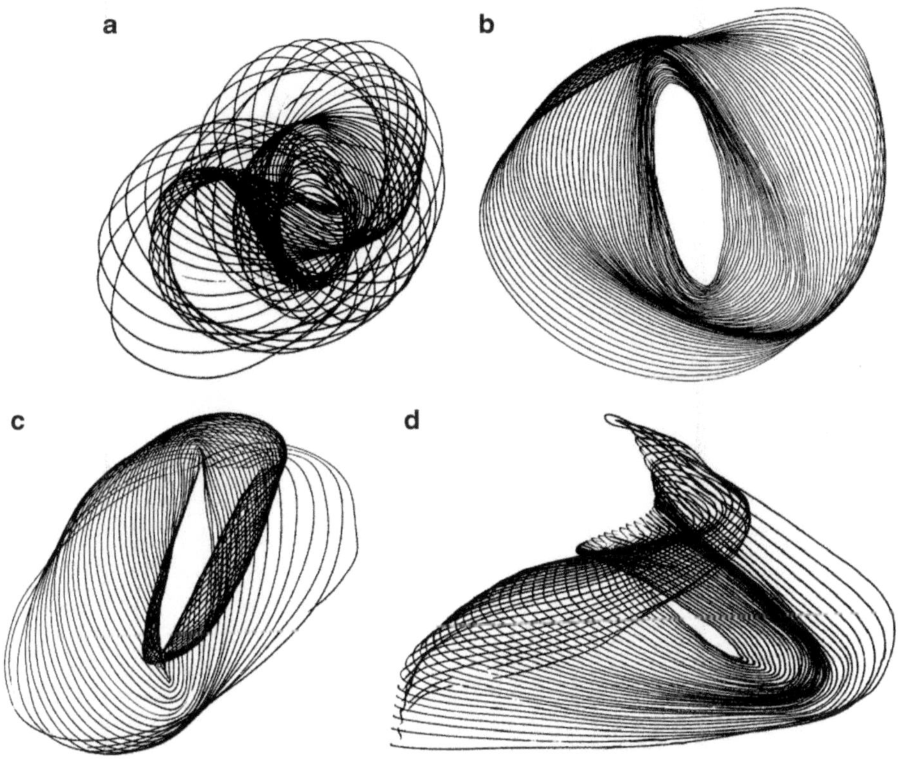

Fig. 8.24 Some harmonograms obtained using the Meccano twin-elliptic harmonograph shown in Fig. 8.5, 1.5 lb (680 g) added to upper bob, 0.5 lb (227 g) added to lower bob

relative frequency, and must also be correctly launched. As an aid to setting up a twin pendulum harmonograph to produce a desired harmonogram Benham (1909) suggests the creation of tables of weight positions versus pendulum period. He also gives guidance on how to adjust a twin-elliptic harmonograph. In practice, it is difficult to reproduce launching conditions exactly, so each harmonograph is different (Cundy and Rollett 1981; Whitaker 2001a). Whitty (1893) used an electromagnetic device to ensure correct launching of his a twin pendulum harmonograph. Hand launching a twin-elliptic harmonograph to give countercurrent rotation is difficult, and practice is needed (Cundy and Rollett 1981). Benham (1909) gives detailed instruction for launching a twin-elliptic harmonograph to give both concurrent and countercurrent rotation.

A twin pendulum harmonograph gives its best results when the two pendulum frequencies are nearly equal, and a twin-elliptic harmonograph gives its best results for high relative frequencies such as 2.5:1 or 3:1 (Cundy and Rollett 1981). Benham (1909) suggests that the twin-elliptic harmonograph is appropriate for relative frequencies of 2:1 to 4:1.

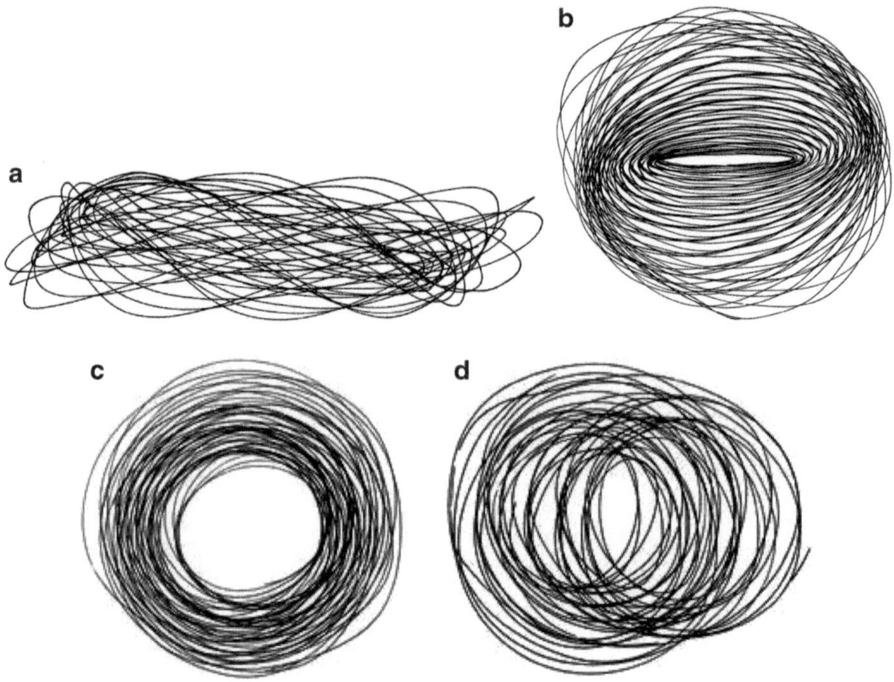

Fig. 8.25 Some harmonograms obtained using the Meccano twin-elliptic harmonograph shown in Fig. 8.5, 0.5 lb (227 g) added to both bobs

References

Anonymous (1930a) Meccano instructions for outfits nos. 4 to 7. Meccano Ltd., Liverpool

Anonymous (1930b) Meccano twin-elliptic harmonograph. Special instructions for building Meccano super models. No. 26. Meccano Ltd., Liverpool

Anonymous (1955) A fascinating designing machine. Meccano Mag 40(7): 384–385

Anonymous (2010a) Lissajous figures. http://www.youtube.com/results?search_query=Lissajous+figures&search_type=&aq=f Accessed 4 Feb 2010

Anonymous (2010b) Lissajous figures. http://physci.kennesaw.edu/javamirror/explrsci/dswmedia/lisajous.htm Accessed 7 Feb 2010

Ashton A (2003) Harmonograph. A visual guide to the mathematics of music. Walker & Company, New York

Bazley TS (1875) Index to the geometric chuck: a treatise upon the description, in the lathe, of simple and compound epitrochoidal or "geometric" curves (with plates). Waterlow & Sons, London (Reprinted by Nabu Public Domain Reprints)

Benham CE (1909) Descriptive and practical details as to harmonographs. In: Newton HC (ed) Harmonic vibrations and vibration figures. Newton & Co., London, pp 26–86

Case WB (1996) The pumping of a swing from the standing position. Am J Phys 64(3):215–220

Case WB, Swanson MA (1990) The pumping of a swing from the seated position. Am J Phys 58(5):463–467

Coxeter HSM (1961) Introduction to geometry. Wiley, New York

Cundy HM, Rollett AR (1981) Mathematical models, 3rd edn. Tarquin Publications, Stradbrooke

Falconer K (1990) Fractal geometry. Mathematical foundations and applications. Wiley, Chichester

Goold J (1909) Vibration figures. In: Newton HC (ed) Harmonic vibrations and vibration figures. Newton & Co., London, pp 89–162

Greenslade TB (2010) Instruments for natural philosophy. http://physics.kenyon.edu/EarlyApparatus/index.html Accessed 5 Feb 2010

Kœnig R (1865) Catalogue des appareils d'acoustique construits par Rudolph Kœnig. Rudolph Kœnig, Paris

Lamb H (1923) Dynamics, 2nd edn. Cambridge University Press, Cambridge

Loney SL (1913) An elementary treatise on the dynamics of a particle and of rigid bodies. Cambridge University Press, Cambridge

Newton HC (1909a) Simple harmonographs. In: Newton HC (ed) Harmonic vibrations and vibration figures. Newton & Co., London, pp 1–25

Newton HC (ed) (1909b) Harmonic vibrations and vibration figures. Newton & Co., London

Roura P, González JA (2010) Towards a more realistic description of swing pumping due to the exchange of angular momentum. Eur J Phys 31(5):1195–1207

Stong CL (1965) Zany mechanical devices that draw figures known as harmonograms. Sci Am 212(5):128–130, 132, 134, 136

Tea PL, Falk H (1968) Pumping on a swing. Am J Phys 36:1165–1166

van den Berg H (1997) A modern Meccano harmonograph. Constructor Q 36:18–25

von Seggern DH (1990) CRC handbook of mathematical curves and surfaces. CRC Press, Boca Raton

Whitaker RJ (1991) A note on the Blackburn pendulum. Am J Phys 59(4):330–333

Whitaker RJ (2001a) Harmonographs I. Pendulum design. Am J Phy 69(2):162–173

Whitaker RJ (2001b) Harmonographs II. Circular design. Am J Phy 69(2):174–183

Whitaker RJ (2005) Types of two-dimensional pendulums and their uses in education. In: Matthews MR, Gauld CF, Stinner A (eds) The pendulum. Scientific, historical and educational perspectives. Springer, Dordrecht, pp 377–391

Whitty HI (1893) The harmonograph. Jarrrold & Sons, Norwich

Yang T, Fang B, Li S, Huang W (2010) Explicit analytical solution of a pendulum with periodically varying length. Eur J Phys 31(5):1089–1096

Index

L.P. Pook, *Understanding Pendulums: A Brief Introduction*, History of Mechanism and Machine Science 12, DOI 10.1007/978-94-007-1415-1,

MIX
Papier aus verantwortungsvollen Quellen
Paper from responsible sources
FSC® C105338

If you have any concerns about our products,
you can contact us on
ProductSafety@springernature.com

In case Publisher is established outside the EU,
the EU authorized representative is:
Springer Nature Customer Service Center GmbH
Europaplatz 3, 69115 Heidelberg, Germany

Printed by Libri Plureos GmbH
in Hamburg, Germany